TROUBLED WATERS: ECONOMIC STRUCTURE, REGULATORY REFORM, AND FISHERIES TRADE

The fishing industries of New England and [barcode: D0223943] rently in a period of deep economic crisis. Th: date analysis of the industries and a definitive new perspective on fisheries trade in general.

The authors show how the Canadian and U.S. fisheries shared common roots but developed into distinct entities, with Atlantic Canada specializing in the production of frozen fish for global markets and New England serving the higher-value fresh-fish markets. They illustrate how public policies have reinforced these differences in market and industrial structure, and argue that, without improved management, neither country's industry will be able to successfully expand into critical new markets. Overall, they reveal the tremendous impact that economic institutions and public policies have had on this troubled industry.

This is the first study to make a comparison between the fishing industry policies and regulatory structures of Canada and the United States, and to assess their effects on the economic growth of the respective industries and on trade in fresh fish between the two countries. By taking a comparative approach, the authors have created a deeper understanding of both systems and proposed some interesting policies based on regional integration.

PETER B. DOERINGER is a professor in the Department of Economics at Boston University and a visiting scholar at the Center for Business and Government at Harvard's John F. Kennedy School of Government.

DAVID G. TERKLA is an associate professor of economics, chair of the Economics Department, and a member of the Environmental Science Program faculty at the University of Massachusetts.

They are co-authors, with Philip Moss, of *The New England Fishing Economy: Jobs, Income and Kinship* and, with Gregory Topakian, of *Invisible Factors in Local Economic Development*.

PETER B. DOERINGER AND
DAVID G. TERKLA

Troubled Waters: Economic Structure, Regulatory Reform, and Fisheries Trade

UNIVERSITY OF TORONTO PRESS
Toronto Buffalo London

© University of Toronto Press Incorporated 1995
Toronto Buffalo London
Printed in Canada

ISBN 0-8020-0683-3 (cloth)
ISBN 0-8020-7639-4 (paper)

Printed on acid-free paper

Canadian Cataloguing in Publication Data

Doeringer, Peter B.
  Troubled waters : economic structure, regulatory reform
  and fisheries trade

  ISBN 0-8020-0683-3 (bound)   ISBN 0-8020-7639-4 (pbk.)

  1. Fish trade – Atlantic Provinces. 2. Fish trade –
  New England. 3. Fisheries – Economic aspects –
  Atlantic Provinces. 4. Fisheries – Economic aspects –
  New England. 5. Fisheries – Government policy –
  Atlantic Provinces. 6. Fisheries – Government policy
  – New England. I. Terkla, David G. II. Title

  HD9464.C23A74 1995   338.3'727'09715   C95-930998-5

University of Toronto Press acknowledges the financial assistance to its
publishing program of the Canada Council and the Ontario Arts Council

Contents

7
Conclusions for Policy 138

Tables and Figures

FIGURES

Abbreviations

AGAC	Atlantic Groundfish Advisory Committee
AGAP	Atlantic Groundfish Adjustment Program
CAFSAC	Canadian Atlantic Fisheries Scientific Advisory Committee
CCF	Capital Construction Fund
CPI	Consumer Price Index
DFO	Department of Fisheries and Oceans
DRIE	Department of Regional Industrial Expansion
FPI	Fisheries Products International
FRCC	Fisheries Resource Conservation Council
FVOG	Federal Vessel Obligation Guarantee
GDP	Gross Domestic Product
ICNAF	International Commission for the Northwest Atlantic Fisheries
ITQ	individual transferable quota
MEY	maximum economic yield
MSY	maximum sustainable yield
NAFO	North Atlantic Fishery Organization
NCARP	Northern Cod Adjustment and Recovery Program
NEFMC	New England Fishery Management Council
NMFS	National Marine Fisheries Service
NOAA	National Oceanic and Atmospheric Administration
OSY	optimum sustainable yield
OY	optimum yield
TAC	total allowable catch
TAGS	The Atlantic Groundfish Strategy
UI	unemployment insurance
USDOC	United States Department of Commerce
USDOL	United States Department of Labor
USITC	United States International Trade Commission

Preface

This study of the New England and Canadian fishing industries has proved to be a longer and more challenging intellectual journey than we had anticipated. Our interest in the industry began with a request in 1982 from the U.S. Department of State to assist its 'Socioeconomic Task Force' in determining the consequences for the New England fishing industry and the regional economy if there were to be an adverse decision by the World Court in a U.S.–Canadian fishing boundary dispute. This introduction to the industry opened a window into a complex world of labor market and production relationships, which caused us to rethink many of our assumptions about the way in which fisheries labor and capital were likely to respond to economic shocks.

In the aftermath of the 1984 World Court decision that divided the Georges Bank fishing grounds between the United States and Canada, we extended our research, with the generous support of the William H. Donner Foundation, to include the study of industry adjustment in Atlantic Canada. Our particular emphasis was on finding opportunities for cooperative relationships between the industries in the two countries that would be to their mutual advantage in the short term and that would eventually contribute to the diversification of their coastal economies. This phase of our research resulted in the publication of a book, *The New England Fishing Economy* (with Philip I. Moss), the development of counterpart materials on the fisheries industry in Atlantic Canada, and the beginnings of an 'institutional' model of fisheries trade flows between the two countries.

The third stage of our project has involved a detailed study of the economic institutions of the industry and of the public policies that have affected its evolution in both countries during the 1980s and 1990s. With research support from the Canadian government's Canadian Studies Institutional Research Program, we were able to update our earlier findings and to pursue issues arising from the recent expansion of Canadian fresh-fish exports to the United States. The decline of groundfish stocks in the early 1990s provided a test of our earlier hypotheses and required the updating of our findings.

Following the industry for over a decade has allowed us the luxury of examining how its economic structure has evolved in response both to the boom and bust in landings and to secular changes in market organization and production arrangements. We have observed a shift in the focus of policy in the two countries from the territorial division of fishing grounds to the measurement of the effect of industry subsidies on trade and then to the fundamentals of common-property resource regulation. We believe our study analyzes these policy issues from a new perspective and reaches some important new conclusions about the interaction between market forces and economic institutions and how this interaction affects international competitiveness.

A project of this length and scope would not have been possible without the assistance of individuals and organizations, too numerous to name, that have given unsparingly of their time and resources to help to educate us in the mysteries of the industry. Our field research and interviews depended on the cooperation of many busy people in both countries, and we hope we have faithfully reported their contributions.

We were fortunate in being joined by Philip Moss, Susan Peterson, and Sean Nolan during various parts of the research program, and we were ably assisted by a number of students at Boston University (including Leonard Schneck, Ricardo Bitran, Sonia Brunschwig, Shelley Drowns, Jeremy Berndt, and Audrey Watson), who played major roles in the field research and data analysis. Ms Kim Little conducted a number of retailing interviews for us, and Ronald Butt, a student at the University of Massachusetts Boston, assisted with the data collection and computer programming for the fisheries trade model. Professors Daniel Georgianna and William Hogan of the University of Massachusetts Dartmouth generously shared with us their data on fisheries processing and landings. Charlene Arzigian shepherded the project through its early development, Kim Little took on major editorial and administrative responsibilities in the preparation of manuscript drafts, and Laura D'Amore saw the project through to completion.

Tony Verga of Gloucester, Howard Nickerson of the Offshore Mariners' Association, Linda Howlett of the Seafood Workers Union, and

Edward J. McLeod of the National Marine Fisheries Service (NMFS) were valuable sources of information on the industry. Joan Palmer helped us with NMFS data. Michael McCarthy of the Food Marketing Institute provided us with information about retail fish sales in the United States. Martin Levine of the Massachusetts Department of Employment and Training worked with us throughout the entire decade.

Martha MacDonald, Patricia Connelly, Leigh Mazany, and Ralph Bannister were of particular assistance in sharing their research on labor markets, trade, and fisheries regulation in Atlantic Canada. We also benefited from a number of conferences organized by the Gorsebrook Institute at St Mary's University and Dalhousie University in Nova Scotia and by the New England Governors' Conference. Without the help of the Department of Regional Industrial Expansion in Nova Scotia and the Department of Fisheries and Oceans we would not have been able to conduct our industry interviews, and we are grateful in particular to Tim Hsu, Janice Raymond, and Leo Brander for helping us with unpublished data of various kinds.

Maureen Yeadon, Frank Fraser, and Walter Martell of National Sea Products were helpful in arranging interviews with corporate officials and providing unpublished data on the industry. Roger Stirling of the Seafood Producers' Association of Nova Scotia gave us similar assistance in arranging industry interviews. Insights from Brian Giroux of the Nova Scotia Groundfish Draggers Association, Garth Dalton of the Independent Seafood Producers of Nova Scotia, and John Kearney of the Maritime Fishermen's Union were of enormous value to the project. Rob Gorham of the Halifax *Chronicle Herald* provided a journalist's perspective on the industry in Nova Scotia. Michael Rooney and Daniel Caron of the Consulate of Canada in Boston provided us with valuable information on the industry at several key points in our study.

In a project of this scope, there are numerous opportunities for error and misinterpretation. We were unusually fortunate in having a number of these problems identified through careful readings of the manuscript by Martha MacDonald, Leigh Mazany, Gene Barrett, Joel Dirlam, Ralph Bannister, and F.M. Scherer. Those errors in fact or interpretation that remain are solely our responsibility.

We gratefully acknowledge the financial support of the William H. Donner Foundation for the early work on this study and of the Canadian Embassy in the United States for its ongoing support of the study. This book was published with financial support from the Jacob

Wertheim Research Fellowship for the Betterment of Industrial Relationships, Harvard University, and the Canadian Embassy in the United States. We would particularly like to thank Dr Norman London of the Canadian Embassy in Washington, DC, for his continued faith in the project, despite the fact that it took longer to complete than we had originally anticipated. None of our sponsors necessarily endorses the findings and conclusions of the study, and we remain responsible for any shortcomings in the final result.

Peter B. Doeringer
David G. Terkla
Boston, Massachusetts
1995

TROUBLED WATERS: ECONOMIC STRUCTURE, REGULATORY REFORM, AND FISHERIES TRADE

1

Efficiency and Institutions

Groundfishing is one of the oldest industries in New England and Atlantic Canada, harvesting species, such as cod and haddock, that are caught near the ocean floor. Like other mature industries, groundfishing has declined in relative importance as the regional economies in both countries have diversified. Nevertheless, the industry continues to be an important source of jobs and income in many coastal regions and port economies.

Far more significant than its economic importance, however, are the insights about competitiveness provided by the interplay among public policies and the economic institutions of the groundfish industry. The industry encompasses both large-scale and small-scale enterprises, traditional and modern technologies, atomistic and vertically integrated industrial structures, and capitalist and kinship forms of labor relations. Each of these institutions imparts a unique identity to the behavior of its sector of the industry, yet all producers are in competition for a common resource and all produce for a common set of markets. This experience with competitive diversity provides the most important lessons from the industry.

The institutional diversity of the industry is matched by sharp differences in the conduct of public policy. The fishing industry is a textbook example of the 'tragedy of the commons,' in which unrestricted access to a productive resource results in its being overexploited. Both the United States and Canada have long recognized this problem and have nominally embraced similar policy goals to restrict overfishing. The United States, however, has followed primarily a laissez-faire and decentralized policy involving a considerable measure of self-management by the industry, whereas Canada has a more tightly controlled and centralized policy that is set by federal government officials. As the stock collapses of the 1990s revealed, however, both approaches have resulted in serious overfishing.

At several critical junctures, Canada has also tailored its regulatory policy to the industrial structure of the industry, and it has coordinated its regulatory policies with other fisheries industrial policies. The most dramatic example of such coordinated policy intervention was in the early 1980s, when Canada combined enterprise catch quotas for large processors with an industry restructuring program that aggressively fostered a large-scale vertically integrated harvesting and processing sector. In contrast, the U.S. regulatory policy has been developed independent of other policy initiatives, and the U.S. government has made no attempt to shape the structure of processing and harvesting. As a result, the New England industry has retained a much more atomistic structure without significant vertical integration.

These differences in economic institutions and economic policies reflect a series of adaptations to common patterns of change in the industries' technological, resource, and market environments. New England and Atlantic Canada, therefore, provide a set of distinct comparisons for studying a variety of issues – how common-property resources should be regulated, the extent to which industrial structure affects international trade, the relative efficiency of large- and small-scale enterprise, and the effect of workplace organization on labor productivity – that are central to current policy debates over economic structure and performance.

CRISIS AND CONFLICT IN INSTITUTIONS

Institutional change in the industry almost always emerges from periods of crisis and conflict. The rapid increase in the number of foreign trawlers during the 1960s, for example, dropped the U.S. share of harvests on the lucrative Georges Bank fishing grounds from almost 90 percent in the early 1960s to a little over 10 percent a decade later and cut total U.S. landings by about 30 percent (Doeringer, Moss, and Terkla 1986a). Atlantic Canada experienced similar problems with foreign fishing, and this crisis led both countries to extend their economic jurisdictions to 200 miles and to install new regulatory arrangements.

The departure of foreign harvesters from within the 200-mile zone by the late 1970s drew additional domestic harvesting effort and processing capacity into the industry in both countries. New vessels were acquired, harvesting technologies were improved, and processing facilities were modernized in the late 1970s and early 1980s. Much of this expansion was uncontrolled and resulted in considerable excess harvesting and processing capacity in both countries (DFO 1993c; Anthony 1990).

The demands of excess capacity led to conflicts over resources between the two countries. One manifestation of this problem was the contest, finally resolved by a World Court decision in 1984, over whether the United States or Canada would control the rich scallop and groundfish resources in Georges Bank and the Gulf of Maine, where the 200-mile economic jurisdictions of both countries overlapped. Although the Georges Bank boundary dispute ended in a draw, significant conflicts have continued over who will dominate the lucrative market for fresh groundfish in the United States.

Canada had been a major exporter of frozen groundfish to the United States for several decades, but Canadian exports were not historically a significant factor in the higher-value-added fresh groundfish markets where most of the New England catch was sold. Beginning in the late 1970s, however, the composition of Canadian exports shifted sharply as the quantity of fresh Canadian cod and haddock exports to the United States almost tripled from 1977 to 1990. While Canadian fresh groundfish exports were increasing, New England landings were falling, and Canadian exports equaled over one-quarter of U.S. production by the mid-1980s.

This penetration of New England's traditional markets for fresh groundfish had a damping effect on fresh-fish prices in the United States. This price effect adversely influenced the economic prospects of New England fishermen, who depended on rising prices to offset the gradual declines in catch. In the face of this crisis of new head-to-head competition in fresh-fish markets, the New England industry repeatedly blamed public policy in Canada for unfairly subsidizing fisheries production. In the mid-1980s the New England industry finally sought tariff protection against fresh Canadian fish (USITC 1984).

Canada countered that its policies were not intended to subsidize the industry but were directed at achieving long-term efficiencies in a 'common property' industry that had developed considerable excess capacity. In the Canadian view, the United States had also subsidized its industry but was primarily at fault for not achieving comparable regulatory efficiency. The United States International Trade Commission (USITC) adjudicated this dispute, awarding a minimal countervailing duty of less than 6 percent on fresh whole groundfish and rejecting any duty on fresh groundfish fillets. Contrary to much of the rhetoric about unfair subsidies, this ruling suggests that the net effect on trade of differential subsidies in the two countries has not been large.

In general, however, the warning signs of booming capacity and intensifying harvesting pressures were largely ignored, because catch values and incomes were rising. As a result, the industry in both countries now faces an unprecedented crisis of stock depletion. The emergency measures taken to restore stocks have prompted a number of short-term policies to shore up falling incomes and remedy job loss in the industry. In the longer term, however, both countries recognize that a smaller and very different industry must emerge from this crisis and that substantial reforms in regulatory practices must occur.

THEORIES OF ECONOMIC PERFORMANCE

The past two decades of crisis and conflict have substantially altered the structure of the industry. The shift in the pattern of fresh-fish trade between Atlantic Canada and the United States that has occurred during this same period points to the possibility that these structural changes may have affected the relative performance of the industries in the two countries.

Trade shifts are normally indicative of changes in relative costs and prices. However, none of the usual cost and price explanations seems to hold. Technological changes have been similar in both countries; the timing of the trade shift is not consistent with the depreciation of the Canadian dollar; and the USITC decision seems to rule out the industrial-subsidy explanation. Lower transportation costs remain a plausible explanation for some of the increased trade in whole fish between southwest Nova Scotia and New England, but they would not explain the increase in fresh-fillet trade outside the New England region.

The puzzle posed by these trade patterns presents an excellent opportunity for examining different theories of economic performance and competitive advantage. Mainstream neoclassical theory, for example, emphasizes the role of efficient resource allocation as a means of maximizing economic welfare and provides the analytical basis for regulation of common-property fisheries resources. According to neoclassical theory, public policy and other economic institutions become important only when there are market failures resulting from market externalities or information imperfections. Institutional theories, on the other hand, assign a major role to market failures and see institutions as central to efficient production. Radical institutional theories place less emphasis on competitive efficiency and are concerned more with distributional issues surrounding excess profits and other quasi-rents.

Common Property Market Failures

The fishing industry is a classic example of market failure arising from 'common property' problems in the harvesting sector (Anderson 1986). It is a well-established principle in resource economics that, without private ownership of fisheries stocks, profit maximization and unrestrained competition among firms will lead to inefficient over-exploitation of the resource. Achieving efficient catch level, therefore, requires regulatory institutions that restrict fishing effort.

Economic theory defines the efficient level of catch as the point where the net social benefits of harvesting are maximized. Both New England and Atlantic Canada have developed regulatory policies that are at least nominally guided by this common property efficiency standard. Each country, however, has adopted very different regulatory practices to achieve this goal. Atlantic Canada relies on a wide range of tools, including entry limitation and a quota system to control catch, while regulation in New England has been limited largely to gear restrictions and the closing of fishing areas.

The current collapse of groundfish stocks in New England and Atlantic Canada has led to claims that the regulatory processes in both countries have been equally defective. While alternative regulatory models are needed for improving long-term industrial performance in both New England and Atlantic Canada, the most recent stock crisis should not overshadow the constructive policy lessons of the last two decades that are provided by comparisons between the two countries.

As this study will show, the causes of the stock collapse are somewhat different in New England than they are in Atlantic Canada. The declines in New England stocks are almost solely attributable to domestic overfishing, while faulty stock analyses, changes in the oceanic environment, and overfishing of transboundary stocks by foreign trawlers join overfishing by the domestic fleet as significant causes of the stock collapse in Canada. The problem of ineffective regulation of overfishing in both countries, however, has similar origins in the small-scale sector of the industry.

Market Failures and Industrial Structure

Although the groundfishing industry in both countries involves numerous small vessels and processors, in Atlantic Canada there is also a vertically integrated sector built around large and hierarchical processing firms. This large-scale sector accounts for roughly half of the volume of groundfish landings in the region.

The effect of differences in industrial structure on the regulation of market failures in the fishing industry is rarely addressed, because the 'common property' literature generally presumes competition among atomistic producers. Industrial structure is often linked, however, to a different set of market failures, which arise out of imperfections in information and inefficiencies in market transactions (Williamson 1975; Stiglitz 1991).

Vertical integration of production, for example, may be more efficient than 'arm's length' competitive market relationships, when the information needed to control quality or to coordinate production decisions through atomistic markets is costly (Demsetz 1988). Similarly, the hierarchical structure and bureaucratic rules of large firms can be efficient substitutes for market transactions when information imperfections and principal-agent problems make market-mediated transactions unusually costly (Williamson 1975; Lazear 1991; Katz 1986).

The fishing industry is characterized by both serious information imperfections and economic relationships that are conducive to principal-agent problems. For example, the most successful vessels depend upon hard-to-measure skills of the captain and hard-to-monitor skills of fishermen. Product quality is also difficult to observe, and, despite the atomistic character of the industry, there are 'small numbers' problems, because product perishability limits the ability of buyers and sellers to search widely for transaction opportunities.

Finally, there may be differences in the sources of efficiency between large- and small-scale producers. The efficiency advantages of small-scale harvesters and processors often lie in their inherent flexibility with respect to gear and species, whereas the large-scale sector may be able to tap various types of scale economies.

Institutional Theories of Industrial Structure

Institutional theories of industrial structure echo many of these same themes of market failure, but they also attribute institutional arrangements to the social norms that emerge within economic organizations and to considerations of power as well as efficiency. For example, behavioral theories of the firm stress organizational over economic incentives for promoting the productivity of managers, and recent versions of efficiency wage theory have emphasized the possibility of resolving market failures through altruistic relationships between managers and their employees (Simon 1991; Akerlof 1984).

Other studies stress the importance of institutions as a source of innovations and superior economic performance (Peters and Waterman 1982; Chandler 1962; Chandler and Daems 1980; Porter 1980). Organizational innovations such as the simplification of skilled jobs and the development of 'high-commitment' human resources practices are examples of institutional arrangements that can contribute to business performance.

Some theories stress institutional inertia and bureaucratic rigidity (Olsen 1988; Weber 1964). According to this view, rigidities tend to perpetuate inefficient institutions over long periods of time, even when they are in head-to-head competition with firms operating under more efficient institutional arrangements (Elbaum 1986; Elbaum and Lazonick 1986; Lorenz and Wilkinson 1986). This belief has led some analysts to claim that government subsidies were needed to compensate for the bureaucratic inefficiencies of these large firms (Barrett and Davis 1984).

The evolution of industrial structure in the groundfish industry bears considerable resemblance to the processes characterized by institutional theories. The initial growth of large-scale, vertically integrated frozen-fish processors in Atlantic Canada is partly a story of business innovation in response to changing market and technological opportunities. On the other hand, the continued consolidation and concentration of large-scale production was accomplished by government fiat, rather than by private initiative. After this restructuring, the balance of production between the large- and small-scale sectors has been defined by governmental regulatory allocations, rather than by market competition. Another example of institutions is the prevalence of kinship employment practices in the small-scale sector that have an important influence on employment and earnings.

On balance, the institutional differences between atomistic and large-scale, vertically integrated producers seem to be more closely correlated with the political economy of public policy and the social demography of the workforce than with imperfections in information. After a decade of industrial restructuring, however, there remains considerable debate over the economic efficiency of large-scale, vertically integrated production structures compared with that of small-scale production.

Radical Political Economy

A third set of explanations of institutions focuses on the power of large-scale enterprises in market economies. This theory posits a tendency for large-scale enterprises to monopolize market economies. 'Monopoly capital' is thought to wield both political and economic power to increase profitability by restricting product market competition at the expense of small producers and to reduce wages, simplify work, and otherwise shift the balance of power and control at the workplace from workers to managers (Baran and Sweezy 1966; Braverman 1974).

Regulatory Rents
Radical theory sees regulatory policies as reflecting the politics of distributing the economic rents created by restrictions on catch. The 'efficient' regulation of common-property resources generates artificial 'scarcity' rents, and the distribution of these rents is controlled by the types of regulatory instruments selected to limit resource exploitation.

   Mainstream theories of economic institutions provide guidelines for efficient regulatory practices and for measuring rent-seeking by vested economic interests (Krueger 1974). They are not well-suited, however, for analyzing the interplay between common property 'efficiency rents' and 'rent-seeking' behavior that is common to fisheries regulation in both countries. For example, area closures and gear limitations in New England have complicated effects on fishermen who specialize in different species. Inshore fishermen with fixed gear may prefer regulations that close fishing grounds because this restriction reduces the damage to their gear by trawlers, while trawler owners would prefer to keep fishing grounds open and to rely on controls of gear to limit catch. Similarly, the allocation of quotas in Atlantic Canada between the captive fleets of large processors and the small-scale sector of independent fishermen is a major factor governing the distribution of fisheries revenue between the two sectors.

Industrial Crises
Radical theories also suggest that economic institutions serve to balance the competing objectives of different groups within an economy – labor, small-scale producers, and large corporations. This 'balance wheel' framework predicts that institutional changes will often be preceded by periods of institutional stability during which the accumulated effects of changing markets and technology lead to a growing mismatch between institutional configurations and economic efficiency (Boyer 1988).

As this mismatch grows, economic performance deteriorates and the tensions over the distribution of income increase. Small mismatches are associated with minor economic problems and are typically resolved through incremental institutional corrections. Large mismatches, however, can result in severe economic crises, which can be corrected only by major structural reforms that realign institutions and markets.

A number of analysts have examined the Canadian fishing industry from this perspective (Apostle and Barrett 1992a). They have argued that production by small-scale, independent fishermen and processors is the most efficient economic structure for the industry because of its highly seasonal and uncertain patterns of catch. The reasoning behind this conclusion is that small-scale producers have the motivation to regulate catch efficiently because of the strong attachment to fishing and to port communities of the labor force in the small-scale sector. Such attachments create both social and economic incentives to protect the fisheries resources from overharvesting. Small-scale vessels also have the added advantage of flexibility, which allows them to switch gear and harvest the most efficient mix of species.

In contrast, the large-scale harvesting and processing industry is inherently inefficient. It suffers from high fixed costs and bureaucratic rigidities, which outweigh any scale economies; it uses political power to acquire subsidies to enhance profits at the public expense; it secures larger quotas than are socially desirable to increase its market share; and it harvests these quotas using destructive methods that excessively deplete fish stocks.

Periodic crises, such as those caused by the current collapse of northern cod stocks in Atlantic Canada, are predicted by radical theories. The collapse of fishing stocks recurrently threatens the viability of the 'inefficient' large-enterprise sector, and continued public intervention has been necessary to sustain the large corporate firms. Regulatory adjustments and industry subsidies have been used to bail out troubled firms during minor crises, while enterprise quotas and the policy-induced restructuring of the Atlantic Canadian offshore processing industry illustrate more radical policy changes designed to deal with major economic crises.

A MIXED APPROACH

None of these theories fully explains the contrasts in economic institutions and economic performance between the groundfishing industries of New England and Atlantic Canada. Each one, however, illuminates important dimensions of the structure and performance of the industry.

Since many fisheries products are standardized commodities and the structure of the industry includes many small firms that actively compete with one another, mainstream economics has much to contribute to the understanding of the market incentives in the industry. Neoclassical economics also highlights the importance of information imperfections and market failures in understanding the regulation of the industry.

At the same time, there are significant elements of increasing returns, organizational learning, and social norms that also affect the efficiency of the industry. They are best explained by institutional theories of firms and labor markets.

Finally, common-property regulation necessarily generates quasi-rents when it reduces overfishing; various subsidy arrangements can do so as well. The industry has a history of both regulation and subsidies and of active political institutions – lobbying groups, trade unions, and large corporations – that are evidence of the kind of rent-seeking behavior to which radical theories of political economy may best apply.

Accordingly, this study draws upon a mix of theories in analyzing the efficiency and distributional implications of the fishing industry's economic institutions. These theories are combined with quantitative evidence and extensive case-study materials to provide a thorough assessment of the factors governing the economic performance of the industry. The result is a perspective on the industry in which economic institutions are the driving forces behind efficiency, competitiveness, and distributional outcomes.

THE STORY IN BRIEF

In this study the strengths and weaknesses of two contrasting production systems in the industry are examined. One is the easy-entry system of small harvesters and processors, where production and distribution are coordinated through competitive markets. The other is a large-scale, vertically integrated system, where production and distribution are coordinated within a management hierarchy. The small-scale system has advantages of flexibility in gear and choice of species; the large-scale sector is less flexible, but it can achieve economies of scale in production and distribution that are unavailable to the atomistic system and it can target its fishing effort more precisely.

Various combinations of small- and large-scale production and distribution are possible, but the atomistic system corresponds roughly to that of New England and that of the inshore sector of the industry in Atlantic Canada. The large-scale system is found exclusively in the offshore sector of the Canadian industry. Both production systems operate in a common-property environment with large stochastic fluctuations in stock abundance. Long-term efficiency implies regulation to limit landings, but the regulatory challenges are different under the two systems.

Ease of entry and the large numbers of harvesters and processors make the monitoring and control effort costly in the atomistic sector. Moreover, the flexibility and diversity of this sector argue for a decentralized regulatory process that can recognize the possibility of social and community-based regulation of common-property problems and can accommodate the efficiencies associated with the species and gear flexibility of small-scale producers.

In contrast, the barriers to entry and the concentration of producers that are characteristic of the large-scale sector make it easier to observe and control fishing effort. Because the large-scale sector is also less flexible than the small-scale sector, the regulatory process can be more centralized and more tightly linked to species-specific stock assessments.

The decentralized and flexible policy system corresponds roughly to the arrangements traditionally followed in New England, and the centrally controlled system most closely resembles Canadian regulation of the large-scale corporate sector. The experience of New England and Atlantic Canada provides a powerful test of the efficacy of interventionist policies compared with more laissez-faire policies. It also offers predictions for which set of production arrangements and labor market institutions is most likely to prevail.

Choosing the balance between these systems, however, is not simply a matter of comparing efficiency properties and net social benefits of each system at a particular time. The atomistic policy system, because of its flexibility, will dominate in an industry that is highly volatile and prone to sudden and unexpected collapses in catch. In contrast, the large-scale, vertically integrated system has efficiency advantages if the underlying biology of the industry can be stabilized at or near the level required for efficient harvesting, as defined by neoclassical common-property theory. The allocation of catch between these systems, therefore, is a matter more of the quality of biological science and the political will of government to grant property rights to the most efficient producers than of the choice among different economic theories of the industry.

Until recently, the direction of efficient policy in both countries seemed clear. The atomistic system of small-scale producers had the advantage in supplying fresh whole fish for white-tablecloth markets of specialty stores and restaurants in the United States. Because it was costly to regulate, regulatory constraints were de facto more flexible in both countries, and this flexibility was reinforced in New England by the decentralized system of self-management by the industry.

In the long term, however, it appeared that the large-scale system in Canada would come to supply the bulk of the fresh-fish market in the United States. Canadian landings were roughly double those of New England, and this resource base gave Canada the clear numerical advantage. Given these resources, the relatively steady landings of Canadian groundfish and the regulatory policy of enterprise quotas favored the scale-economies of the large-scale sector. Furthermore, this sector had a competitive advantage in opening highly profitable mass markets for fresh fish in supermarket chains.

Supermarkets require large supplies of high-quality processed fish that can be contracted for weeks in advance at known prices and quantities. The large-scale sector had developed a mass-production and marketing capability of proven competitiveness in global markets for frozen fish, and this mass-production and distribution system also could be used for fresh-fish fillets.

Prior to the late 1970s, differences in freshness and quality were the major factors that channeled Canadian fish into relatively low-value-added frozen-fish markets instead of higher-value-added fresh-fish markets. As a result, the New England industry had a protected market, which was experiencing a strong growth in demand and rising prices. Thereafter, Canada became increasingly competitive in U.S. fresh-fish markets.

The initial growth in Canadian exports of fresh groundfish was stimulated partly by cheaper and faster transportation for whole fresh fish from southwest Nova Scotia to New England distribution centers, partly by higher prices for fresh fish that compensated for transportation costs, and partly by the relatively greater abundance of stocks. The USITC findings also suggest that a small export advantage in whole fish may have been generated by government subsidies.

The growth in exports of fresh fillets, however, cannot be so easily explained. Fresh fillets were not a subsidized product and were sold mainly to midwestern markets, where there was no particular benefit from cheaper transportation to New England. Instead, the growth in fillet trade reflected the advantages of a regulatory policy that permitted targeted fishing and restructuring policies that improved the productive efficiency of the large-scale sector. It also represented the increasing ability of large-scale firms to secure landings of sufficient quality for fresh markets. With better-quality fish, the large-scale sector was able to build upon its frozen-fish marketing and distribution capabilities to open new markets for fresh fillets.

These developments suggested that Canadian supplies of whole fish would continue to restrain price increases in white-tablecloth markets, while Canadian fresh-fish fillets would dominate the emerging supermarket demand. The loosely regulated New England industry would be squeezed between a less and less abundant resource and falling catch levels, which would not be fully compensated by rising prices in white-tablecloth markets, and it would lack the mass-production capability to compete with Atlantic Canada for supermarket sales.

THE FUTURE OF THE INDUSTRY

The recent stock collapse in both countries has sidetracked these predictions about the future development of the fishing industry, at least for the time being. Crises and conflict, however, have always been a stimulus for change in the institutions of the industry. The current crises – depleted groundfish stocks, moratoriums on harvesting, and displacement of labor and capital from the industry – have highlighted the contradictions of rising capacity and efficient resource management and are leading to a substantial reassessment of the industry and its policies in both countries.

Rather than supporting an exclusive reliance on the traditional economics of common-property regulation, or rejecting it outright in favor of community-based self-regulation, this study strongly demonstrates the need for blending standard common-property regulation with industrial and labor market policies that recognize the structure of the industry's economic institutions. Such a broad-gauge approach to policy is needed to address the tensions between policies directed at two different types of market failure – one that is based upon the failures of large-scale, vertically integrated market structures and the other upon failures of atomistic markets – and between the efficiency and distributional consequences of regulatory policy.

The need for broad-gauge policy coordination applies with equal force in the international sphere. If the fisheries stocks shared by the United States and Canada are to be managed efficiently and if common

fisheries markets are to be fully developed, there must be policy harmonization as well as free trade between the two countries.

The important issues for the future, however, reach beyond trade and
policy coordination. They include a vision of a smaller and more stable
industry, political decisions about which sectors will gain and which
will lose through downsizing, what arrangements should be used to
alleviate the employment and income consequences of catch fluctuations, and how New England and Atlantic Canada will share established and emerging markets for fresh fish. By defining the institutional,
as well as the economic, parameters of efficiency and by documenting
the key distributional concerns, the chapters that follow offer a blueprint
for constructive change in the industry.

PLAN OF THE BOOK

The study begins with an overview of the main economic features of the
industry in the postwar period. In chapter 3, the changing economic
structure of the industry is examined. Regulatory and industrial polices
are discussed in chapter 4. Issues of labor market structure, productivity,
and income distribution are analyzed in chapter 5. The consequences of
changing economic structure and public policy interventions for international competitiveness and fisheries trade are addressed in chapter 6.
Future directions for public policy are explored in the concluding
chapter. Detailed statistics on the industry in New England and Atlantic
Canada are provided in the appendix (tables A.1–A.16).

2

An Overview of the North Atlantic Fishing Industry

Alantic Canada dominates the Canadian saltwater fisheries, accounting for 75 percent of landings by weight and 68 percent by value in 1993 (DFO unpublished preliminary data 1993).[1] In contrast, New England represents a smaller part of the U.S. saltwater fisheries, providing only about 6 percent of the tonnage and a little over 16 percent of the value of landings nationwide in 1993 (USDOC, *Fisheries of the United States*, 1993). The New England fleet, however, harvests almost all of the nation's Atlantic groundfish. It accounted for 88 percent of the groundfish tonnage in 1990 (USDOC 1990) and for around 60 percent by weight and value (1991) of U.S. sea scallop landings (USDOC, *Fisheries of the United States*, 1993; Georgianna, Dirlam, and Townsend 1993).

The Canadian industry is much larger than its U.S. counterpart, reflecting its larger fishery resource base and its role as a major world exporter of frozen and salted fish. The Atlantic Canada fleet harvests three to four times the total tonnage of the New England fleet and two to ten times the tonnage of haddock, cod, and flounder (figures 2.1, 2.2; tables A.1 and A.2). The unit value of the groundfish and scallop catch, however, is considerably less than that in New England because of differences in the mix of species and the proportion of catch destined for high-value fresh-fish markets (tables A.3, A.4).

Over the years, the fishing industries in New England and Atlantic Canada have been marked by frequent changes in regulations, markets, and technologies. Each has also specialized in different product lines and has periodically faced major declines, only to return to prosperity as new markets and new technologies have emerged. Although both industries are currently facing crises because of sharp declines in groundfish stocks, history suggests that even this downturn will prove to be only temporary.

FIGURE 2.1

Landings, all species, Atlantic Canada and New England: 1977–93

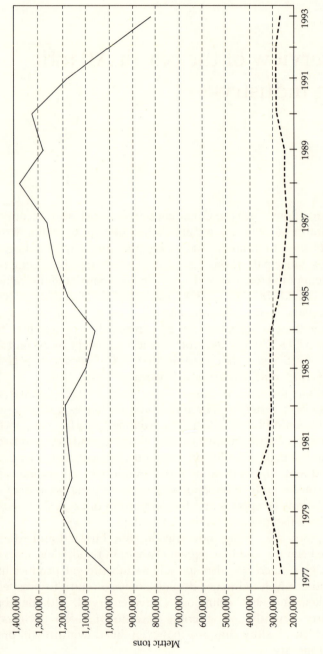

NOTES

New England data: Landings in metric tons of live weight, except for univalve and bivalve molluscs, for which weights are exclusive of their shells. Source: USDOC, various years.

Atlantic Canada data: Landings in metric tons of live weight (fish and shellfish only). Sources: DFO, various years, for 1977–87; preliminary data from DFO for 1988–93.

FIGURE 2.2
Landings by species, Atlantic Canada and New England: 1977–93

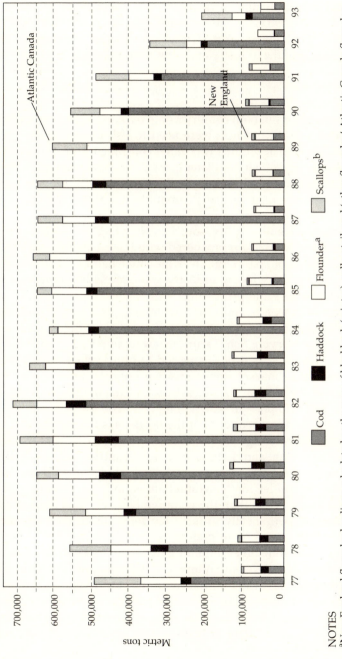

NOTES
[a]New England flounder landings calculated as the sum of blackback (winter), yellowtail, and 'other flounder.' Atlantic Canada flounder = 'small flatfishes.'
[b]Data on New England scallop landings not available for 1992 and 1993.
Atlantic Canada data: Landings in metric tons of live weight. Sources: DFO, various years, for 1977–87; preliminary data from DFO for 1988–93.
New England data: Landings in metric tons of live weight, except for scallops, for which weights are exclusive of their shells. Sources: USDOC, various years; Georgianna et al, 1993.

The future level and value of catch will depend critically on the performance of the industry, regulatory practices, and other public policies that affect competitiveness. This chapter provides an overview of the political economy of the industry and how it has evolved.

BACKGROUND

The fishing industries in both countries date back to the colonial period, when wooden vessels and nets were used to fish for cod (Innis 1940; Barrett 1992a). The final product was either consumed fresh or salted for later use. While the saltfish (and later canned fish) were exported, fresh fish were consumed only locally in both countries until the late nineteenth century, when relatively small amounts of iced fish were shipped by rail to inland cities.

In the early nineteenth century, southern New England ports (such as New Bedford and Nantucket) prospered with the expansion of whaling and the processing of sperm oil. Over the latter half of the nineteenth century, however, the whaling industry was gradually supplanted by groundfish harvesting and processing.

In the late nineteenth and early twentieth centuries, steamships replaced schooners and refrigeration supplemented salting and canning as a method of preserving fish. Tariffs also were imposed by the United States during the latter part of this period to protect the growing New England groundfish market from Canadian imports (Innis 1940). One consequence of the U.S. tariffs was a large exodus of Canadian fishermen to New England ports in search of higher incomes.[2]

A more important consequence of the tariffs was the search by Canada for new markets for its fish. The Canadian industry entered markets in the Mediterranean and Latin American countries where saltfish, dried fish, and pickled fish were in great demand. Although these markets provided an outlet for Newfoundland's and Nova Scotia's fish, they yielded lower prices than the lucrative New England market and were also subject periodically to tariff barriers. Throughout the early 1900s, tariff barriers and trade preferences continued to shift – sometimes favoring and sometimes inhibiting Canadian exports (Innis 1940).

The industry also underwent a number of technological changes in processing and marketing in the late nineteenth and early twentieth centuries that helped to open new markets. The first major frozen-fish companies were founded in the United States in the early twentieth century, and innovations, such as mechanical filleting and rapid-freezing processes (invented in 1923), expanded markets for groundfish. These new technologies, particularly the increased use of refrigeration and freezing, required large capital investments and eventually led to increased industrial concentration in the frozen-fish processing industry (Innis 1940).

The perishability of fresh fish, however, meant that this market largely remained confined to coastal areas. It took the development of refrigerated trucking in the 1930s to open inland markets, and by 1940 almost 60 percent of Boston's fresh and frozen fish were shipped over 200 miles. In the late 1940s the U.S. fresh-fish market grew again as air transport began to link Boston with fish markets in major cities across the United States (White 1954). Despite these improvements in transportation technology, the longer distances and poorer transportation networks continued to limit Canadian access to fresh-fish markets in both Canada and the United States. As a result, Canadian companies concentrated on expanding their presence in the growing, worldwide frozen-fish market, leading to neglect of fresh-fish markets throughout the 1960s and 1970s (DFO, *Annual Statistical Review*, 1955–76). As late as 1980 only 11 percent of the key groundfish species were sold as fresh products, while frozen and salted products accounted, respectively, for 66 percent and 23 percent of sales (table A.5).

Because fresh fish had a higher value and was experiencing a rapid growth in demand, the New England industry came to specialize almost exclusively in the production of fresh products. Most of these fresh fish traditionally have been consumed within the New England and greater New York regions, but they are now being shipped to other parts of the United States as well. Meanwhile, the freezing of locally caught fish became a residual market for New England fish that could not be absorbed in the fresh market, and (in contrast to the situation in Canada) frozen processors came to rely almost solely upon Canadian and other foreign sources of frozen-fish blocks (White 1954).

These long-standing differences in access to the growing U.S. fresh-fish market had implications for the prosperity of the industries in the two countries. Frozen and traditional saltfish markets can tolerate lower-quality fish than fresh markets can. Transportation times and shelf life are less relevant to these markets, and physical defects can be trimmed away when fish are processed into large, pressed blocks. Premium prices, however, are paid for high-quality fish that are suitable for fresh markets. The highest prices are commanded by fish that are caught and landed on the same day and are plump, firm, and free of physical defects.

Canada, therefore, aspired to shift production out of lower-value frozen markets into fresh markets. Aided by reduced transportation costs, new technologies that allowed for mechanized filleting, and packaging improvements that extended the shelf life of fresh fish, Atlantic Canada (particularly Nova Scotia) began to increase its share of the U.S. fresh-fish market during the 1980s (USITC 1984, 1986).

Industrial Structure and Public Policy

The differences in product specialization by the two countries – Canada in frozen and salted fish and the United States in fresh fish – led to differences in industrial structure in the two countries. Currently, the New England fishing industry is highly atomistic. Vessels are independently owned and operated, as are fresh-fish processors, and there is little vertical integration between harvesting and processing.

Like New England, Atlantic Canada has an atomistic sector composed of independently owned vessels and small processors. But an equally important sector consists primarily of two large, corporate processing firms and is responsible for most of the frozen-fish processing. This sector is vertically integrated – owning its own fleet of offshore vessels and handling marketing and distribution as well as processing – and accounts for the industry's being much more concentrated in Canada than in New England.

These variations in industrial structure and concentration are caused by differences in public policy, as well as differences in the mix of fish products each country produces. Although both countries regulate their harvesting sectors, the federal and provincial governments in Canada have always been more deeply involved in the structure of the industry than their U.S. counterparts have been (Sinclair 1987; Apostle and Barrett 1992a). In Canada, for example, fisheries regulation has been tailored to accommodate differences between the atomistic and the large-scale sectors, and the government has intervened directly in the restructuring and refinancing of the largest processors.

The restructuring of the industry was intended to bring both increased efficiency to production of frozen fish in the large-scale corporate sector and an increased capacity to sell fresh fish in the United States, particularly to mass retail markets where large supplies of product at predictable prices are important elements of competitiveness. The relative efficiency of the atomistic and corporate sectors remains controversial, however, and, while the corporate sector has developed a core of supermarket customers in the United States, the atomistic sector still accounts for the largest share of fresh-fish exports to the United States.

ECONOMIC DEVELOPMENT

The fishing industry is central to the economy of Atlantic Canada, where it accounts for about 15 percent of GDP in the major fishing provinces of Nova Scotia and Newfoundland (DFO, *Annual Statistical Review*, various years). Although once a mainstay of the New England economy, the industry today contributes only a tiny fraction to regional GDP. It remains an important source of jobs and income in only a handful of large ports and in some rural coastal regions.

Processing and harvesting accounted for almost 16 percent of Newfoundland employment in 1990 and over 6 percent of employment in Nova Scotia (DFO 1993c). In that year, there were almost 16,000 full- and part-time registered fishermen in Nova Scotia and almost 29,000 in Newfoundland (figure 2.3; table A.6; DFO 1993c). The corresponding figures for processing employment (1988) are over 7,000 in Nova Scotia and over 11,000 in Newfoundland (figure 2.4; table A.7). The industry is highly seasonal, however, so that approximately half of the fishermen and a substantial fraction of the processing labor force in both provinces are part time.[3]

The New England fleet employs only about 6,500 fishermen (1985), some 3,500 of whom are offshore fishermen with substantial year-round employment (Doeringer, Moss, and Terkla 1986a). While there are variations among states, the fishing industry in Massachusetts (where most of the fleet is concentrated) accounts for less that 1 percent of statewide employment, although the fraction is much higher in the largest ports, Gloucester and New Bedford.

Employment has been contracting in recent years. In Massachusetts, for example, the year-round harvesting labor force was about 1,500 in 1992, down from around 2,000 in 1990 and over 3,500 in 1980, and there was a combined frozen and fresh processing workforce of around 2,700 in 1992, down from almost 3,200 in 1990 and over 5,700 in 1980 (figures 2.3, 2.4; tables A.6, A.7).

The importance of the fishing industry to specific port economies has created pressures in both countries for various subsidies to the industry.

FIGURE 2.3
Number of employees in fish harvesting in Massachusetts, Nova Scotia, and Newfoundland 1977–93

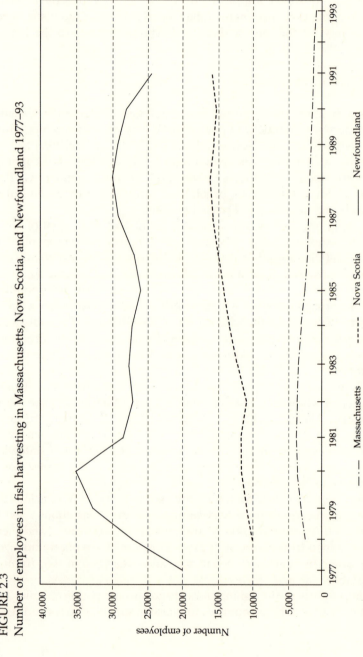

Massachusetts data: Source: USDOL, various years, for SIC 0912 (finfish).
Nova Scotia and Newfoundland data: Sources: DFO, various years, for 1977–87; unpublished data from DFO for 1988–91.

FIGURE 2.4
Number of employees in fish processing plants in Massachusetts, Nova Scotia, and Newfoundland: 1977–92

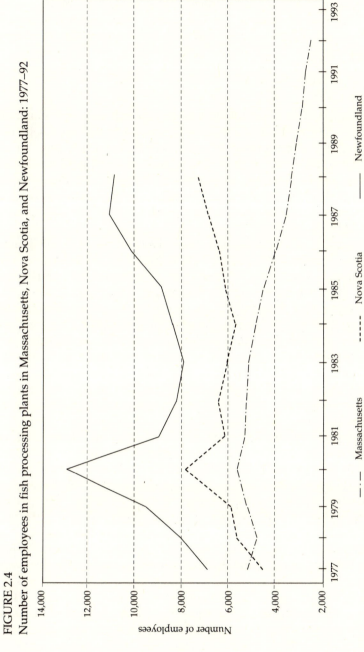

Massachusetts data: Source: USDOC, various years.
Nova Scotia and Newfoundland data: Sources: DFO, various years, for 1977–87; unpublished data from DFO for 1988.

These economic development policies have historically played a relatively larger role in Atlantic Canada than in New England because in the former the industry is larger and there are fewer alternative employment opportunities.

New England Port Economies

The fishing fleet in New England is distributed unevenly along the coast, over half of the fleet being concentrated in Massachusetts and a substantial portion located in Maine (USDOC 1986a). Most of the groundfish (72 percent of the tonnage and value in 1990) are landed in Massachusetts, followed by Maine (15 percent), and Rhode Island (8 percent) (USDOC 1991b). The Massachusetts fleet also accounts for over 80 percent of the total value of scallop landings (Georgianna, Dirlam, and Townsend 1993).

The largest New England ports – New Bedford (Mass.) and Gloucester (Mass.), Portland (Maine), and Point Judith (R.I.) – handle around 75 percent of the total New England catch value and about two-thirds of its groundfish processing activity (Doeringer, Moss, and Terkla 1986a; NEFMC 1993, 190). Altogether, these ports have over 400 offshore vessels, more than half of New England's offshore fleet, and fishing is a year-round and relatively high-wage activity. There are slightly fewer than 600 processing workers in New Bedford and about half that number in Gloucester, where frozen-fish processing is also important (Georgianna, Dirlam, and Townsend 1993). Portland is also becoming a distribution center as a result of its new 'display' auction and improved fishing pier.

Medium- and small-sized ports are scattered up and down the New England coast, ranging in size from Provincetown (with a year-round population of 3,500 and a fleet of about fifty groundfish vessels) to tiny ports in Maine with only two or three vessels intermittently engaged in a mixture of shellfishing, lobstering, and groundfishing. The smallest New England ports tend to have only inshore boats with crews of two to four. These boats are limited to day trips and most fish less than 150 days per year. Low catch levels and highly seasonal production generally result in low incomes and preclude the development of a processing sector in these ports.

Some of the large ports have diversified economies. The New Bedford area has a substantial manufacturing sector, although much of it consists of relatively low-paying industries such as apparel and food processing, and Portland is a rapidly growing center for transportation, banking, and commerce. In many smaller ports, however, there are often few employment alternatives to tourism and fishing, particularly along the rural coast of eastern Maine.

Port Economies in Atlantic Canada

The bulk of the Canadian North Atlantic fishing fleet is located in Newfoundland (55 percent) and Nova Scotia (21 percent), with smaller concentrations of vessels in Quebec (10 percent) and New Brunswick (10 percent). Newfoundland and Nova Scotia dominate the industry, accounting for 78 percent of all fish landings, 77 percent of the value of the catch, 75 percent of processing employment, 70 percent of all registered fishermen, and 59 percent of all registered fish plants (DFO, *Annual Statistical Review*, 1991).

There are some large, offshore vessel ports in Atlantic Canada that are similar to Gloucester and New Bedford in terms of the concentration of processing facilities, prevalence of year-round work, and relatively high incomes. Lunenberg (N.S.), a community of about 4,000 located fifty miles from Halifax, is the largest of these ports. It is home port to about 900 fishermen and is the 'flagship' port for National Sea Products, which has the largest processing plant in the area (Raymond 1985). Lunenberg also has a marine sales and repair sector that serves much of southwest Nova Scotia (Pross and Heber 1982). Altogether, fishing and processing in Lunenberg County provide about 1,700 full-time equivalent jobs and account for about 10 percent of the county's labor force (Hache 1989). Lunenberg has achieved a certain degree of economic diversity, with manufacturing accounting for about 40 percent of employment and the business and personal services sector providing another 25 percent (Raymond 1985).

Atlantic Canada is also dotted with many small and medium-sized ports, almost all of which are fisheries dependent. For example, of the 364 ports in Nova Scotia, all but six had populations of 2,500 or less, and over 80 percent had populations under 500 in 1976 (Kirby 1982). A typical medium-sized port might have a population of about 650, with half or more of the labor force employed in highly seasonal fishing and processing (Raymond 1985).

FISHERIES LABOR MARKETS

Labor markets in the industry are essentially local. Crew in New England, and on independent vessels in Atlantic Canada, are recruited from among family and friends, as is much of the labor supply for Canada's offshore sector (Jorion 1982; Ilcan 1985; Doeringer, Moss, and Terkla 1986a; Squires 1990). Processing also draws primarily from local labor markets, except for the largest and highest wage processors, which draw labor from a wider area.

Few fishing skills are transferable to other occupations and, with the exception of some part-year crew on inshore vessels, fishermen tend to have little work experience outside the fishing industry. Those fishermen with outside work experience have usually been employed in other primary-product industries, such as forestry or farming, or sometimes in tourism-related industries (Doeringer, Moss, and Terkla 1986a, 1986b).

Employment and Earnings

Fishermen's earnings are based on the revenue of their vessels, and incomes, at least in the short term, are closely tied to changes in the value of catch. In periods of rising catch or rising prices, a fisherman's income can increase substantially above that of alternative onshore employment. In the longer term, however, entry of fishermen and vessels has meant that earnings have tended to approximate those of semi-skilled onshore jobs.

For example, after the establishment of the 200-mile economic zone, the landed value of all species in Atlantic Canada rose 48 percent in real terms between 1977 and 1979 and by 24 percent in New England (figure 2.5; table A.8). Within a year harvesting employment began to rise in both countries. Between 1978 and 1980 employment in harvesting rose 11 percent in Nova Scotia, by 33 percent in Newfoundland, and by 29 percent in Massachusetts (figure 2.3; table A.6). While the correlation between employment and catch is most evident in periods of strong growth in catch revenue, subsequent employment trends have been consistent with fluctuations in the prosperity of the industry in the two countries.

Earnings data for individual fishermen are hard to obtain and extraordinarily unreliable. Nevertheless, the pattern of reported earnings on larger vessels in Massachusetts is consistent with these revenue and employment shifts (table A.9). The real earnings of fishermen were relatively high in 1979 and then declined more or less steadily (apart from the boom year of 1987, when they rose by 23 percent) as catch and revenue fell. Similar patterns of earnings peaking in 1987 and then declining are apparent in the Canadian data for the late 1980s (table A.9).

FIGURE 2.5
Real landed values, all species, Atlantic Canada and New England: 1977–93

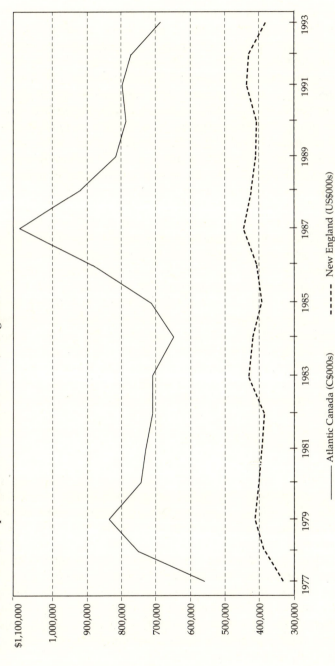

——— Atlantic Canada (C$000s)   - - - - - New England (US$000s)

NOTES

Atlantic Canada data: Landed values in C$000s. Real values calculated using Canadian CPI, 1986 = 100. Sources: DFO, various years, 1977–87; preliminary data from DFO for 1988–93.

New England data: Landed values in US$000s. Real values calculated using US CPI, 1982–4 = 100. Source: USDOC, various years.

Offshore fishermen tend to work year round and to be relatively better paid than inshore fishermen, who work fewer weeks a year. Distinctions between inshore and offshore earnings are particularly pronounced in Atlantic Canada, where surveys show that earnings of offshore crew are over twice those of inshore fishermen (Thiessen and Davis 1988). Comparable survey data are not available for New England, but interviews in Gloucester and New Bedford during the mid-1980s, as well as estimates by the National Marine Fisheries Service of *full-time* earnings, indicate that incomes of offshore fishermen averaged around $25,000, considerably above those of inshore fishermen in smaller ports and well above those of full-time workers in manufacturing (Hogan, Georgianna, and Huff 1991; Doeringer, Moss, and Terkla 1986b).

Processing employment is semi-skilled and fluctuates with the level of landings. The increase in landings after the 200-mile limit was established drove processing employment up by 64 percent in Nova Scotia and by 86 percent in Newfoundland between 1977 and 1980 (figure 2.4; table A.7). Massachusetts processing employment rose by only 8 percent between 1977 and 1980 (figure 2.4; table A.7), but catch-driven changes in employment in the fresh-fish sector are presumably masked by the greater employment stability in the larger frozen-fish processing sector, which does not depend on domestic landings. Case studies repeatedly find that processing wages are much lower than those for harvesting and are tied to the local labor market rather than to changes in the value of the catch (Doeringer, Moss, and Terkla 1986a; Macdonald and Connelly 1986b).

Internal Labor Markets

The most striking feature of fisheries labor markets, however, is the diversity of institutional structures that govern internal labor market relationships. The informal organizational practices of entrepreneurial capitalism, the bureaucratic practices of corporate capitalism, and the familial practices of kinship capitalism operate side by side in both countries (Doeringer, Moss, and Terkla 1986b; Barber 1992; Ilcan 1985; Jorion 1982; Squires 1990; MacDonald and Connelly 1986b).

To some degree, this diversity in internal labor market structures reflects elements of the industrial structure of the industry, such as the scale of enterprise and the type of ownership. Kinship arrangements, for example, are most common in small-scale, family-owned enterprises, whereas bureaucratic employment practices are more characteristic of the large-scale corporate sector.

Differences in scale and ownership between New England and Atlantic Canada, therefore, result in different mixes of labor market institutions in the two countries. For example, the institutions of corporate capitalism are more prevalent in Canadian than in New England processing because of Canada's large corporate processing sector, whereas the ethnic ownership of offshore vessels in New Bedford and Gloucester results in a greater concentration of kinship capitalism in the New England offshore fleet than is found in Canada's offshore fleet.

MANAGEMENT OF THE NORTH ATLANTIC FISHERIES

The fish-harvesting sector represents a classic case of the common property resource management problem, and efforts to regulate the fishery can be traced to the turn of the century in both New England and Atlantic Canada. Not until after the Second World War, however, when technological innovations such as on-board processing facilities threatened the vast fisheries resource in the Gulf of Maine, was formal regulation attempted. A new regulatory body, the International Commission for the Northwest Atlantic Fisheries (ICNAF), was created to regulate harvesting of the northwest Atlantic offshore stocks beyond the twelve-mile territorial limits of the United States and Canada. ICNAF started with eleven countries as signatories, including Great Britain and the Soviet Union, but it was dominated by the United States and Canada.

Regulation under ICNAF did not encounter serious problems until the late 1960s and early 1970s, when foreign fishing fleets from eastern and western Europe sent large factory vessels into waters off the North American coast. These vessels were capable of harvesting and freezing 25 percent to 50 percent more fish per hour than were traditional vessels (Warner 1983). ICNAF proved ineffective in controlling these harvesting efforts, and the fishing industries in New England and Atlantic Canada faced severe economic losses as stocks declined.

Total catch of all species in New England fell almost 30 percent from its 1965 level to 226,000 metric tons in 1975, and the Atlantic Canada fleet suffered an even larger decline in catch (down 36 percent from its 1968 peak of 1.3 million metric tons). For both New England and Atlantic Canada, however, the effects of declining catch were somewhat blunted by significant increases in the prices of fresh and frozen fish as a result of growing global demand. Between 1965 and 1975 the value of the New England catch rose steadily from $75 million to $154 million, a real increase of almost 20 percent (Doeringer, Moss, and Terkla 1986a). Likewise, the nominal value of the Atlantic Canadian catch increased by almost two-thirds, from C$116 million, to C$191 million between 1968 and 1975, a real increase of 7.2 percent (DFO, *Annual Statistical Review*, various years).

Responding to Foreign Fleets

In 1976, coincident with an international movement to extend sovereign state control over contiguous coastal resources, the United States declared separate governance rights over resources within 200 miles of its coastline to stem the further loss of catch to foreign vessels. Canada followed suit in early 1977.

The establishment of 200-mile exclusive economic zones marked the beginning of divergent fisheries management policies in the United States and Canada. Two years after the 200-mile exclusive economic zones were established, both countries had adopted their own distinct regulatory systems and ICNAF had been dissolved. Canada opted for the centralized federal management of its fisheries resources, but the United States took the opposite tack, giving substantial authority to regional fisheries councils. Canada also was instrumental in creating the North Atlantic Fishery Organization (NAFO), an international body for managing fishery resources beyond 200-mile national jurisdictions, while the United States withdrew from participation in the international management of northwest Atlantic fish stocks.

The new management regimes in both countries were based on the shared belief that heavy foreign fishing was the primary cause of the poor state of the stocks (Acheson 1984). With the exclusion of most foreign fishing, landings in the United States and Canada did increase dramatically in the late 1970s. Between 1977 and 1982, for example, total landings rose by 19 percent in Atlantic Canada, and New England landings peaked in 1980 at 36 percent above their 1977 level (figure 2.1; table A.1). It has since been determined, however, that there were also several good year classes that contributed substantially to the rebounding of stocks independent of the changing management structures (Acheson 1984).

While catch levels were rising, the unit values (in real terms) of all species remained relatively stable in both countries as increases in supply were offset by substantial increases in demand (figure 2.6; tables A.3,

FIGURE 2.6
Real unit values, all species, Atlantic Canada and New England: 1977–93

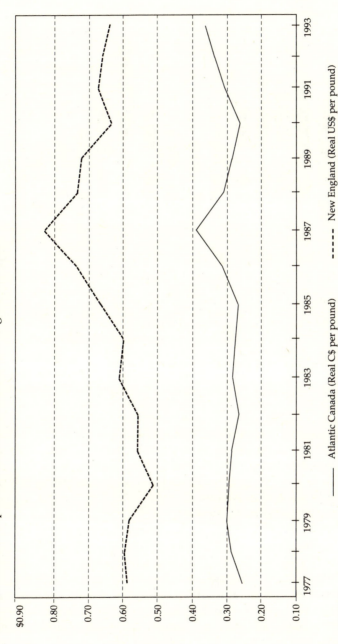

NOTES

Atlantic Canada data: Prices calculated in C$ per pound of live weight from landings and values taken from DFO, various years, for 1977–87, and preliminary data from DFO for 1988–93. Real values calculated using Canadian consumer price index, 1986 = 100.
New England data: Prices calculated in US$ per pound of live weight from landings and values taken from USDOC, various years. Real values calculated using U.S. consumer price index, 1982–4 = 100.

——— Atlantic Canada (Real C$ per pound)   ----- New England (Real US$ per pound)

FIGURE 2.7

Real landed values by species, Atlantic Canada: 1977–93

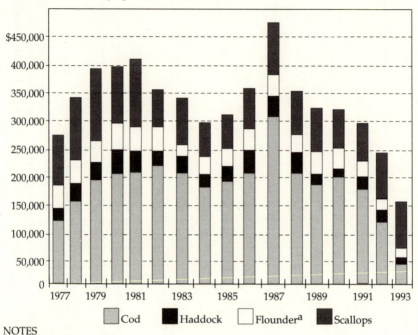

NOTES

Landed values in C$000s. Real values calculated using Canadian CPI, 1986 = 100.
[a]Small flatfishes.

Sources: DFO, various years, for 1977–87; preliminary data from DFO for 1988–93.

A.4). This growth in output, even with stable prices, drove up the real value of landings. Real values of catch in Atlantic Canada rose by 25 percent between 1977 and 1982 and by 16 percent in New England (figure 2.5; table A.8). For key species such as cod, the real value of landings rose by 92 percent in Atlantic Canada and by 37 percent in New England during this period (figures 2.7, 2.8; table A.10).

Recurrent Catch Declines

Since this initial surge in landings, neither country has been able to realize its expectations of a healthy fishery that could sustain the increased landings. After 1980 New England's total catch began to fall and by 1987 was 31 percent below its 1980 high (figure 2.1; table A.1). Cod landings fell and then began to recover, but a particularly dramatic and persistent

FIGURE 2.8

Real landed values by species, New England: 1977–93

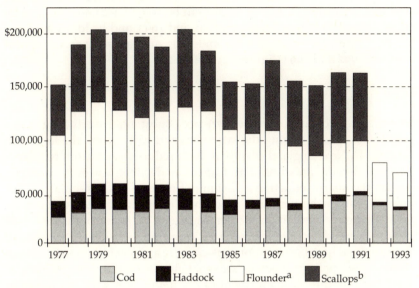

NOTES

Landed values in US$000s. Real values calculated using U.S. CPI, 1982–4 = 100.

[a]Flounder landings calculated as the sum of blackback (winter), yellowtail, and 'other flounder' landings.

[b]Information on scallop landings not available for 1992 and 1993.

Sources: USDOC, various years; Georgianna, Dirlam, and Townsend 1993.

fall occurred in other key groundfish species such as haddock and flounder (figure 2.2; table A.2).

As landings fell, rising demand and buoyant prices initially sustained industry revenues. The real value of the cod, haddock, and flounder catch in New England, for example, fluctuated around an average of $147 million between 1980 and 1987. By the early 1990s, however, price increases were no longer offsetting continued declines in groundfish stocks. The real value of landings of these species declined by almost 37 percent between 1987 and 1993, almost three-fourths of this decline occurring after 1991 (figure 2.8; table A.10).

During the 1980s the fisheries resource remained relatively stronger in Atlantic Canada than it was in New England. Catch levels (all species) continued to rise, although erratically, throughout the 1980s, and the real value of catch almost doubled between 1977 and its peak year of

1987 (figure 2.7; table A.10). As was the case in New England, however, there were particularly sharp declines in key groundfish species. Cod landings, for example, fell by 39 percent between 1982 and 1991 and by an additional 77 percent between 1991 and 1993, and haddock landings fell by 53 percent between 1981 and 1991 and by another 40 percent between 1992 and 1993 (figure 2.2; table A.2).

Regulatory Policy

Although both countries have experienced substantial overfishing in key groundfish species during the late 1980s and early 1990s, the fall-off in landings in New England mainly reflects reduced catch from relatively constant fishing pressure. The Canadian decline, however, partially reflects the sharp reduction in fishing effort due to the regulatory closures of many fishing areas. Canada has followed a limited-entry management policy and in particular segments of some industries uses market-oriented techniques, such as transferable quotas. In contrast, New England has chosen a more laissez-faire approach – relying more on 'open-access' policies (such as seasonal closures and mesh-size restrictions) than on direct management of catch and fleet size. The radically different approaches to management of the stocks taken by the two countries also may partially explain the relatively better Canadian experience during the 1980s.

The Canadian approach to management is also much more centralized and, at least on the surface, is somewhat insulated from direct political pressures from fishermen and processors. The regional council system adopted in the United States, in which the different fishery interest groups are represented, has been more vulnerable to pressures from private interests to allow some overfishing than has the Canadian regulatory system. The Canadians have not been immune to such pressures, however, as is revealed by the use of regulatory policy and government subsidies to support the fishing industry in the many underdeveloped areas of Newfoundland and Nova Scotia where fishing and processing are the primary sources of employment (Copes 1983; Charles 1992).

Declining stocks have resulted in a new crisis for the industry in both countries. Even though unusually large year classes of cod and yellowtail flounder helped to sustain New England landings in 1989 and 1990, the underlying stocks are now in poor shape (Massachusetts Task Force 1990; Anthony 1990). Their condition has sparked lawsuits from environmental groups and intense political pressure to improve fisheries management, which in turn have resulted in regulations limiting entry into the fishery for the first time in its history and significant closures on Georges Bank. In Canada, depleted stocks have forced a cut in 1994 Canadian quota levels for major groundfish species by almost 75 percent since 1989 (DFO 1994). Reduced quotas were followed by the closure of the northern cod fishery off Newfoundland after July 1992 and the closure of most other groundfishing areas since 1993.

Summary

The fishing industry in New England and Atlantic Canada has had a volatile history in terms of its economic fortunes. Products and markets have shifted, technology has changed in both harvesting and processing, international competition has become more intense, government regulation has increased, and direct and indirect subsidy programs have been used to strengthen the industry in both countries.

The economic structure of the industry is quite different in both countries. The Canadian industry has both a highly concentrated sector of large-scale, vertically integrated firms and an atomistic harvesting and processing sector. This dual structure has been reinforced by industrial policies, subsidies, and regulatory policies. In contrast, the New England industry is atomistic and government policy has been relatively laissez-faire. These distinctions in industrial structure also spill over into the structures of labor markets in the two countries.

Both industries experienced substantial booms following the introduction of their respective 200-mile economic zones, but this prosperity has given way to diminishing landings and declines in income as regulatory policy has failed to stem excessive exploitation of stocks. These problems are emerging at a time when head-to-head competition between the two industries over ownership of stocks and relative shares of the highly lucrative fresh-fish market in the United States has also been increasing.

The combination of stock declines and increased competition has raised the level of economic conflict between the two countries and has highlighted the differences in their industrial and regulatory policies. The industries in the two countries are now at a crossroads in terms of which system of industrial structure and public policy will prevail.

3

Industry Structure and Market Performance

Although Canada and the United States harvest similar species from the northeast Atlantic using similar technologies, the industrial structures of their harvesting and processing industries have evolved quite differently. The Canadian industry is much larger and, in general, much more concentrated than that of New England. While the New England industry is centered almost entirely around production for the fresh-fish market, the vast bulk of the Canadian harvest is destined for frozen-fish markets.

In this chapter, the structure of the harvesting and processing industries in each country will be discussed, followed by a description of the marketing institutions in both countries. The chapter concludes with a discussion of the possible changes in industry structure that might occur in each country as U.S. fresh-fish markets evolve.

THE STRUCTURE OF THE NEW ENGLAND HARVESTING INDUSTRY

The New England fleet has grown substantially over the last two decades, from 703 vessels of five tons or greater in 1965 to 1,362 vessels in 1992 (table 3.1; Doeringer, Terkla, and Moss 1986a). Most of this growth took place in the late 1970s, however, in response to the extension of U.S. jurisdiction over fish resources to 200 miles from shore. Between 1974 and 1981 the number of vessels almost doubled. Although there was some fluctuation during the last decade, the size of the fleet remained substantially unchanged through 1992.

This near constancy in overall fleet size masks a steady increase in the size of vessels in the fleet throughout the 1980s (table 3.1). The largest vessels (150–500 tons) increased by 50 percent over the decade (split

TABLE 3.1
Number of fishing vessels by size in New England: 1980–92

	5–50 tons	50–150 tons	150–500 tons	Total
1980	616	532	168	1,316
1981	623	550	191	1,364
1982	653	533	190	1,376
1983	581	583	193	1,357
1984	611	595	217	1,423
1985	590	554	217	1,361
1986	540	505	209	1,254
1987	631	493	209	1,333
1988	651	499	242	1,392
1989	599	509	247	1,355
1990	598	512	252	1,362
1991	628	469	241	1,338
1992	681	454	227	1,362

Sources: USDOC 1991a, for 1980–9; USDOC 1994, for 1990–2.

almost equally between scallop dredge and otter trawl vessels), while the number of smaller vessels declined slightly, although with considerable year-to-year fluctuations (USDOC 1991a). Since 1990 the number of large vessels has fallen with the decline in fish stocks, and despite the recorded increase, it is likely that the actual number of small vessels has fallen as well.[1]

Inshore versus Offshore Harvesting Capacity

The fleet is divided between inshore and offshore vessels. Inshore vessels (5 to 50 tons) employ two to three crew and in 1989 gross revenues averaged around $60,000 (about 10 percent of that earned by the largest vessels) (USDOC 1991a). Because of their small size, these vessels fish near shore, usually leaving port and returning on the same day with a maximum trip of two to three days. They rarely travel to the more distant and richer offshore grounds (such as Georges Bank), nor do they fish in rough weather, so most of their effort is concentrated during the summer months. Because inshore vessels' trips are short, their fish are fresher when landed and, therefore, are often sold into the premium-quality fresh-fish market.

While inshore vessels represent half of all commercial fishing vessels in New England, they land a much smaller fraction of the groundfish and scallops than do the offshore vessels. Overall, the inshore sector accounted for only around 20 percent of all New England groundfish landings in 1991 (NEFMC 1993, 184). Within New England, the inshore fleet harvested about 2.5 percent of the Massachusetts and Rhode Island groundfish landings (1990) and larger fractions of those of Maine (24 percent) and New Hampshire (39 percent) (USDOC 1990).

Since more competition exists among vessels in inshore areas than in offshore grounds, and there is greater variability in the availability of species over the course of the year, inshore vessels must be highly flexible in terms of species fished for and type of gear employed. For example, inshore vessels account for most of the landings of lobster and under-utilized species such as squid, skate, and ocean pout.

There are 681 vessels greater than 50 tons (1992) in the offshore fleet. The majority of these vessels have their home port in Massachusetts and most are berthed in Gloucester and New Bedford (USITC 1984; USDOC 1986a). Other concentrations of vessels are found in Portland, Boston, and Port Judith (R.I.).

Offshore vessels have crews of six to twelve and regularly travel 200–300 miles to distant fishing grounds, where they fish for up to one week or more before returning to port. Although weekly trips and longer travel times to and from port mean that even the most recently harvested offshore fish are less fresh than those harvested by the inshore fleet, almost all of the offshore landings are sold in the fresh-fish market.

The larger size of offshore vessels permits them to fish in inclement weather and thus to operate year round. Given their large harvesting capacity, it is not surprising that the offshore vessels harvest almost 90 percent of the annual landings of groundfish and scallops in New England (USDOC 1990, and *Commercial Fisheries News*, 1991).

The average size of offshore vessels increased during the 1980s. As a consequence, the harvesting and revenue-generating capacity of the offshore fleet increased considerably during this period.[2]

Ownership and Entry

New England vessels are owned by independent entrepreneurs (many of whom also captain their boats) or by family-based corporations. There are a few examples of vessels owned by outside investors or fish processors, but they are the exceptions in the fleet. Most of the offshore groundfish vessels are now owned and crewed by ethnic families – Italians and some Portuguese in Gloucester and Portuguese in New Bedford – and few families own more than two boats. Offshore scallop vessels are concentrated in New Bedford; they are owned primarily by families of Norwegian descent but are crewed by Portuguese. Because of widespread individual ownership of vessels, industrial concentration in fish harvesting remains very low.

Until mid-1994 there was no policy limiting entry of vessels into the U.S. industry, but the sizable capital expense (approximately $600,000–$750,000 for a new offshore otter trawl vessel to $1 million for a new scallop vessel) served as a significant barrier to entry. The importance of fishing skill, knowledge of the waters, and reputation for success in the industry further limited entry, although advances in electronic fish-finding and navigation technologies have significantly reduced the competitive advantage of many of these traditional skills in recent years.

Markets

While most New England vessels, particularly those in the inshore fleet, rely on one or two home port buyers to purchase their fish, some offshore vessels sell their fish through auctions (Wilson 1980). Although only a small volume of domestically caught fish is sold directly through auctions held in Portland, Boston, and New Bedford, these auctions are very important in determining ex-vessel fish prices.

The Boston auction takes place at 6:00 a.m. each weekday, and buyers bid 'blind' on portions of a particular boat's catch or on fish trucked in from other ports. The New Bedford auction, which handles a larger volume of fish, begins after the Boston auction ends, and buyers must purchase the entire mix of catch on a boat sight unseen and at a single price for each species. Although few fish are landed in Boston and only a few boats are on the New Bedford auction on any given day, bidders have reasonably good estimates of the landings in Gloucester and New Bedford and in the other smaller ports up and down the coast, as well as of the potential Canadian imports to brokers in Boston and inventories of unsold fish (Peterson and Georgianna 1988). The New Bedford auction, for example, is heavily influenced by Boston prices, information about fish landings along the coast, and demand in New York, an important outlet for much of the fish landed in New Bedford.

Focal Point Prices

Despite the small amount of fish actually auctioned, the auction price is a meaningful price guide or 'focal point' up and down the coast.

Demand for fresh fish is relatively stable in the short term, while supply fluctuates daily in terms of species and quantity. In response to these fluctuations in supply, processors set prices and market their product based upon 'spot' market availability. Our interviews indicate that Boston processors are signaling through the small Boston auction what they will pay for fish from Gloucester and the hinterland ports, knowing the approximate supply in each of the main ports, including New Bedford. Boston has maintained this role as a price setter because of the large volume of fish processed there each day.

Processors and wholesalers use the auction prices in different ways as a basis for negotiation with harvesters. In Boston, for example, fishermen are usually paid the auction price unless their fish are of poor quality. In most cases, vessels contain a range of quality and there is further negotiation between fisherman and buyer over final prices.

Similarly, the New Bedford auction price is set for fish that are top quality, and actual prices may be adjusted downward to reflect the quality of delivered fish. If a fisherman is frequently dissatisfied with the post-auction payments of a particular buyer, he can bar that person from bidding for his catch in the future. Although doing so may provide some control against abuse of the post-auction transaction process, it also reduces competition for the boat, thereby potentially lowering the prices it will receive (Peterson and Georgianna 1988).

In Gloucester, where a majority of the Boston area fish are landed, processors and wholesalers usually pay fishermen 3–5 cents per pound below Boston auction prices. Most Gloucester buyers have commitments from Boston processors to buy their fish at the Boston auction price, but the binding nature of these commitments is subject to dispute. It seems that small buyers in Boston often honor their commitments, but large buyers have been known to insist on lower prices, or they will not accept delivery.

In cases where Gloucester buyers are paid less than the Boston auction price, buyers try to pass on the discount through lower prices to the fishermen. This practice is limited, however, because buyers cannot continually underprice other Gloucester buyers and retain their client relationship with fishermen. Nonetheless, the many other informal transactions between buyers and fishermen – loans, dock space, fuel, and ice – allow considerable room for negotiation over actual ex-vessel prices.

In 1986 Portland began a 'display' auction, in which buyers are able to bid on small lots of fish that are displayed in containers. The goal of the display auction is to reduce the uncertainty about the quality of fish and to improve prices for fishermen who take extra efforts to maintain quality. The auction appears to have been successful, since the volume of fish has more than doubled from around 14.5 million pounds in 1988 to over 30 million pounds in 1993 and the number of buyers purchasing seats on the auction doubled from an initial twelve to twenty-four (Portland Fish Exchange 1994). Nevertheless, the vast majority of fish continues to be landed in Massachusetts, and prices continue to be determined largely by the Boston and New Bedford auctions.

Despite the imperfections in the different auction processes, the widespread use of bilateral negotiations, and wide price spreads among processors and vessels, there are many reasons to believe that actual prices paid tend to be influenced primarily by supply and demand conditions in the region. Although there are few buyers within any particular port, the landings in all ports are usually reflected in the prices paid, and the ability of fishermen to land their fish relatively easily in nearby ports limits any oligopsony power among buyers.

THE STRUCTURE OF THE CANADIAN HARVESTING INDUSTRY

The Atlantic Canada fleet (vessels 10 tons and over) is more than five times larger than its counterpart in New England, with almost all of the difference accounted for by small (10–50 ton) vessels (DFO, *Annual Statistical Review*, 1986).[3] As in the case of the United States, the Atlantic Canada fleet has grown dramatically over the last two decades. The number of vessels greater than 10 tons has more than doubled, with much of the growth taking place in the late 1970s (table 3.2). Although fleet size data for the early 1990s are not available, it is likely that vessel numbers have been falling in recent years, as they have in New England, because of the large decline in fish stocks.

Inshore versus Offshore Harvesting Capacity

The distinction between the inshore fleet and the offshore fleet is far more important in Atlantic Canada than it is in New England because it defines a series of differences between the fleets in terms of ownership, regulatory practices, and the types of product markets served. The traditional definition of 'inshore' vessels for management purposes in Canada is 'vessels less than or equal to 100 feet in length' (or less than approximately 170 tons).[4] The inshore fleet (according to this definition) comprised 98 percent of all Canadian vessels in 1986.

TABLE 3.2
Number of registered vessels by size in Atlantic Canada: 1973, 1978–90

	< 10 tons	10–50 tons	50–150 tons	> 150 tons	Total
1973	25,466	2,928	295	250	28,939
1978	25,472	4,649	343	264	30,728
1979	27,141	5,464	379	276	33,260
1980	25,520	5,629	357	266	31,772
1981	24,190	6,056	425	274	30,945
1982	20,880	5,773	406	268	27,327
1983	23,732	6,278	494	271	30,775
1984	23,280	6,289	491	260	30,320
1985	21,258	6,629	679	221	28,787
1986	21,335	6,698	697	215	28,945
1987	22,194	6,840	693	209	29,936
1988	22,499	6,942	731	218	30,390
1989	22,095	6,989	727	203	30,014
1990	21,418	6,879	714	192	29,203

NOTES: Data for Quebec not available for 1980 and 1982. Post-1984 data have been converted from length to weight according to the following scale:

under 35 feet:	< 10 tons
35 to 54.9 feet:	10 to 50 tons
55 to 99.9 feet:	50 to 150 tons
100 feet and over:	> 150 tons

Source: DFO, various years.

This fleet, however, has historically accounted for only around 50–60 percent of the Atlantic Canada groundfish catch (figure 3.1). This share has been kept relatively constant by regulatory policies that have divided catch quotas between the inshore and offshore fleets (Gardner 1988). In the late 1980s there was a slight shift in catch quotas toward the inshore fleet from 53 percent of the 1982 allocation to 58 percent of the 1989 allocation (Hache 1989). This appears to have been a regulatory response to political pressures from traditional inshore fishermen who were facing increased competition from the growing 'jumbo' inshore fleet of vessels (Halliday, Peacock, and Burke 1992).

Recently, Canadian managers have adopted a definition of 'inshore' that corresponds more closely to the U.S. definition, classifying vessels less than 35 feet in length as 'inshore,' vessels 35–65 feet as 'nearshore,' and vessels 65–100 feet as middle-distance or 'midshore' (Kirby 1982; Hache 1989).[5] In 1987 the newly defined inshore fleet (vessels less than 35 feet) represented 74 percent of the total fleet, but accounted for only 18 percent of the groundfish catch. Nearshore vessels accounted for

FIGURE 3.1

Percentage of total groundfish landings, in metric tons, by vessels under 100 feet, Atlantic Canada: 1970–87

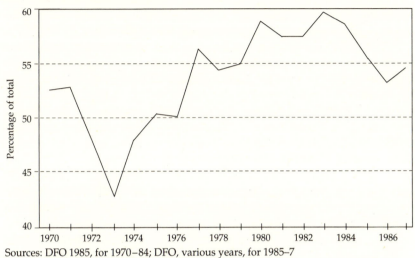

Sources: DFO 1985, for 1970–84; DFO, various years, for 1985–7

24 percent of the fleet and 32 percent of the catch and the midshore fleet for less than 1 percent of the fleet and 3 percent of the catch (DFO, *Annual Statistical Review,* 1987).

In 1987 the offshore fleet consisted of around 209 vessels greater than 150 tons, almost the same number as the U.S. offshore vessels in this size class (tables 3.1, 3.2). A little over half of the offshore fleet is located in Nova Scotia and almost 40 percent in Newfoundland (DFO, *Annual Statistical Review,* 1987). In 1987 the offshore fleet landed over 45 percent of the total Atlantic Canadian groundfish harvest and over 73 percent of the scallops (DFO, *Annual Statistical Review,* 1987). Over four-fifths of the offshore groundfish catch was caught by the largest vessels (greater than 500 tons) in this fleet.

Ownership and Entry

Almost all the inshore, nearshore, and midshore vessels are individually owned and operated, as they are in New England. Unlike the independently owned U.S. offshore fleet, however, most of the offshore groundfish vessels greater than 150 tons are owned by the two major Canadian processing firms. Corporate ownership allows the location and timing of the fleet's fishing activity to be coordinated among vessels, rather than being left to the decision of individual captains. This coordination has enabled the processors to distribute vessel landings among plants along the coast throughout the year in response to prices, catch availability, and considerations of efficient capacity utilization.

In contrast to the situation in New England, most of the growth in the number of vessels in the Canadian fleet during the 1980s was in the medium-sized (50–150 ton) and small-vessel (10–50 ton) sectors, while the number of large (greater than 150-ton) vessels in the Canadian fleet declined. Much of this difference appears to be explained by the effect of restrictions on vessel entry in Canada. Vessels over 65 feet in length (approximately 70–80 tons) have been subject to limited-entry regulations since 1973 (full implementation of limited entry was not in place until 1975). Entry of some smaller vessels has been limited since 1976, and there has been a complete freeze on all new entrants since 1980 (Halliday, Peacock, and Burke 1992). In theory, limited entry means that a new vessel can enter the industry only by purchasing the license from an existing vessel, but there have been problems in implementing these restrictions (MacDonald 1984; Halliday, Peacock, and Burke 1992).

Licensing restrictions (along with the introduction of enterprise quotas in 1983) were effective in limiting the size of the largest-vessel sector (table 3.2), but the number of vessels in the 10–150-ton category grew substantially. The rising landings and value of key groundfish species in Atlantic Canada from 1977 to 1982 (tables A.2, A.10), along with substantial government financial assistance, enabled many small-vessel fishermen to replace their vessels with larger boats. Most of this growth in capacity took place in the 35–65-foot category, since the acquisition of new vessels greater than 65 feet was effectively blocked by the entry restrictions. Overall, the number of vessels 35–45 feet in length increased by 25 percent, and the number in the 45–65-foot class rose by 16 percent (Kirby 1982). Much of the latter increase took place in Nova Scotia, where the number of vessels in the 45–65-foot class increased by almost 45 percent.

This rapid growth in the smaller-vessel fleet resulted in more restrictive entry regulations in 1982 designed to halt the expansion of the 45–65-foot vessels. Consequently, when fish prices increased during the mid-1980s (table A.3), the industry's increased revenue was funneled into a new, smaller (less than 45 feet) class of high-capacity, fiberglass jumbo vessels. In Nova Scotia, the number of jumbo vessels increased by almost one-third between 1985 and 1989, replacing lower-powered vessels in this size class (Hache 1989).

Markets and Prices

Independent fishermen sell their catch primarily to independent processors that produce frozen, salt, or fresh fish. Our interviews revealed that almost all of the Nova Scotia inshore catch is either salted or sold fresh, in contrast to the Newfoundland industry, where almost the entire inshore catch is frozen. Some of the fish caught by the nearshore fleet are also sold into the fresh-fish market, particularly during the winter months, when fresh-fish prices are highest, and the rest is frozen. The midshore vessels are mostly longliners located in Nova Scotia, and their catch is split almost equally between the fresh- and salt-fish markets.

The bulk of the whole fresh fish exported to New England comes from the southwest Nova Scotia inshore fleet, largely because these fish tend to be fresher and more carefully handled than are those caught by the Canadian offshore fleet. Also, many of the small processors that purchase fish from this fleet lack freezing capacity, and therefore whatever fish are not salted are sent into the fresh-fish market.[6]

Prices for this Nova Scotia inshore fleet are set on a landing-by-landing basis and reflect New England prices (Mazany, Barrett, and Apostle 1987). In contrast, almost all of the Newfoundland inshore fleet are part of a marketing 'union' affiliated with the Canadian Auto Workers. Fish prices are negotiated on an annual basis, with the trade association representing the vast majority of the small frozen-fish processors and wholesalers in Newfoundland.

The offshore fleet produces for the large processing companies that own the vessels. Controlling the quality of the catch on such large vessels is difficult because of the length of time at sea required to fill the hold capacity and the sheer quantity of fish brought on board. As a result, much of the Canadian offshore catch is processed for the frozen market, which has lower quality requirements than the fresh-fish market.

The vertical integration of the offshore fleet with onshore processing and distribution allows internal transfer prices, rather than market prices, to be used in the Canadian corporate sector. One set of transfer prices is used to compensate fishermen. These 'crew' prices are set annually and are not subject to short-term fluctuations in supply and demand.

During the 1980s, however, the large processors sought to improve the quality and freshness of their landings and began to market fresh processed fillets to supermarket chains. As part of an attempt to improve the quality of landed fish, quality premiums were introduced into the prices fishermen received. Under National Sea Products' dockside grading program, three grades of fish quality were established, with a 100 percent price differential between the bottom and top grades and a 25 percent spread between the top quality and the middle grade (Gardner 1988).

'External' market prices are also a factor in the offshore industry, because a portion of the fish supplied to the corporate sector comes from direct purchases from independent suppliers and wholesalers. These prices are subject to negotiation, and during the late 1980s and early 1990s National Sea Products sought to expand its purchases from independent fishermen through annual purchasing agreements (Cheticamp 1991, 11). Our recent interviews suggest that this practice is declining.

The Structure of the New England Processing Industry

The fish-processing industry in the United States is sharply divided into fresh and frozen sectors. While a few firms deal in both markets, most specialize in one or the other.

Historically, the frozen processors relied on New England fish. The New England harvest has largely been siphoned off into the higher-value fresh-fish market, however, so that frozen processors now rely almost exclusively on frozen-fish blocks purchased in highly competitive global markets. While frozen processors are largely independent of the rest of the New England fishing industry, they remain in New England because of marketing cachet and, in some cases, to be close to their Canadian owners.

The frozen sector markets almost all of its product in the United States. Its main competitors are Canada and Iceland, but each major producing country specializes in a particular market niche, thus limiting head-to-head competition. U.S. companies dominate the retail market for brand-name frozen fish sticks and portions because of high import duties on frozen processed fish products and the storage and transportation costs of these products, compared with frozen blocks (Kirby 1982). Iceland occupies the higher-quality end of the frozen market, catering to some white-tablecloth restaurants and chains, while Canadian output is concentrated in the lower-end market (large institutional buyers, such as schools and hospitals) for fillets, and in frozen blocks (Kirby 1982).

Fresh-fish processing and distribution is nominally an atomistic industry, and very few processing firms own fishing vessels. Yet high transaction costs, informal relationships between harvesters and processors, and some product differentiation introduce an element of monopolistic competition into the industry.

Scale of Production

Fresh-fish processing is carried out primarily in many small (employing fewer than fifty workers), family-owned and -operated businesses. The scale of output has been relatively stable or declining since the 1970s. The average yearly volume of output per plant in Boston rose 30 percent, from under 1 million pounds during the early 1970s to 1.3 million pounds in 1979, but it has remained about the same through the early 1990s (Georgianna, Dirlam, and Townsend 1993). New Bedford experienced a decline in annual output from an average of 1.8 million pounds in the early 1970s to 1.5 million pounds in 1979 and had fallen to around 1 million pounds by 1991 (Georgianna and Ibara 1983; Georgianna, Dirlam, and Townsend 1993).

Production scale has been constrained by the uncertain availability of domestic fresh-fish landings. Uncertainty of fish supplies and the relatively low markup on processed product have made most processors reluctant to invest in the new automated technologies (such as fish-cutting and -skinning machines) that became available during the 1970s. The high fixed costs of these machines could not initially be supported by the volume of processing, and the subsequent need to remain flexible has limited the economies of scale that can be obtained from growth in plant size in the 1980s and 1990s (Georgianna and Hogan 1986).

Market segmentation and marketing transaction costs also constrain the scale of processing firms. Most processing firms have established market niches by specializing in particular species (flatfish or round-bodied fish) or by catering to a particular clientele – white-tablecloth restaurants, elite fish markets, large-scale catering and food service firms, or wholesalers in large markets such as New York, Philadelphia, or Chicago.

These individual niche markets have grown slowly because the importance of customer loyalty makes developing new clientele costly. Since the shelf life of fresh-fish is short (a maximum of two weeks), it is imperative that processors be able to move their product continually. Customers who frequently shop around greatly increase the uncertainty facing a processor over whether or not he will be able to sell his product quickly, while stable customers enable him to plan his purchases of fresh-fish more efficiently. Since most processors ship their products on a 'buyer approval' basis, it is similarly important for both parties to have a thorough understanding of their product and quality needs in order to avoid frequent haggling over the price of shipments or having processors find another buyer before the product's shelf life expires.

Concern with uncertainty works both ways. Large retailers, such as supermarkets and some restaurants specializing in seafood, also need to be assured of a constant supply of particular fish, and certainty of supply is valued much more than the option to buy at lower prices. Thus, the focus on personal relationships between particular wholesalers and retail buyers (as well as between particular vessels and wholesalers and processors) creates high transactions costs. These costs of switching buyer-seller relationships reinforce the monopolistically competitive nature of the industry and inhibit the consolidation of fresh-fish processing to gain economies of scale in marketing.

Market Structure

Although very few processors own boats, the harvesting industry is more integrated with fish processors than the widespread individual ownership of vessels would suggest. In addition to establishing stable buying and selling arrangements, individual vessels and processors tend to form stable relationships around the provision of supplies and berthing and marketing that further imitate elements of vertical integration.

In Gloucester, and to a lesser extent in New Bedford, vessels rely on processors or wholesalers for fuel and ice and for small informal loans not easily obtainable from local banks. Moreover, until the State Fish Pier was renovated in 1993, there was almost no public docking space in Gloucester, so most boats depended on processors or wholesalers for berthing space.

Beginning in the 1980s, a few processors in different ports sought to strengthen further their ties with fishermen by offering vessels profit-sharing plans or guaranteed premiums for top-quality fish delivered regularly. However, most fishermen continue to prefer purchasing arrangements based on the current price of fish at the time of landing. They feel they can 'beat the odds' by landing their fish on days when prices are higher and thus earn more than they would under a longer-term, guaranteed-price contract with a particular processor.

Geographic Concentration

The northeastern U.S. fresh-fish processing industry is centered in Massachusetts, and there is evidence that the concentration of processing activity in Massachusetts has increased in recent years, particularly in the Boston area (Georgianna, Dirlam, and Townsend 1993). In 1985 Massachusetts plants produced 87.5 million pounds of fresh fillets valued at $184 million (USITC 1986). This figure represented more than 80 percent of all U.S. fresh-groundfish processing (in terms of both quantity and value of fillets), up from 70 percent in 1980 (USITC 1986; Georgianna and Ibara 1983).[7] The number of fresh-groundfish processing plants in Massachusetts has held steady at about fifty throughout the 1970s and most of the 1980s and declined only slightly in 1989 and 1990, while there has been a sharp decline in fish-processing activity in New Hampshire and Maine over the entire period (Georgianna, Dirlam, and Townsend 1993).

Within Massachusetts, about half of the processors are located in Boston and most of the rest are in New Bedford. In recent years, Gloucester fresh-fish processing has practically disappeared, since most fresh-fish processors have converted to wholesaling or frozen processing and almost all fish landed in Gloucester are being processed elsewhere, mostly in Boston (Terkla and Wiggin 1994).

Industrial Concentration

The concentration of production within the fresh-fish processing industry is still quite low. In 1983 (the most recent year for which data are available) the largest producer accounted for a little less than 10 percent of total industry production, and the top four firms accounted for around 28 percent of industry output (USITC 1984). Concentration within particular ports, however, is considerably higher. According to our interviews, the top four firms produced almost 60 percent of the total processed fish in Boston, over 90 percent in Gloucester, and over 80 percent in New Bedford. These concentration ratios correspond roughly to those in 1979 (Georgianna and Dirlam 1982). Buyer concentration is also substantial in some of the more minor ports, such as Plymouth, Provincetown, and Portsmouth.

Despite new entrants in Boston and New Bedford, the dominant firms have remained the same for the last ten to twenty years, and these larger firms are the ones that seem to be surviving in the face of sharp declines in supply in recent years. Their wider supply networks and broader species variety give them a competitive advantage over smaller firms, which cannot guarantee delivery to customers (Georgianna, Dirlam, and Townsend 1993). Moreover, as prices rose and supermarkets began to demand greater predictability of price and supply during the 1980s, these processors diversified their sources of supply to include imports from Canada and elsewhere as well as locally harvested fish.

If the low levels of fish stocks continue to persist, many smaller processors and wholesalers may begin to drop out of the industry. This trend will lead the industry away from its current monopolistically competitive structure toward a somewhat more oligopolistic one.

THE STRUCTURE OF THE CANADIAN PROCESSING INDUSTRY

The current Canadian processing industry is far larger and more diverse than the New England industry, and its markets are more global. In 1987 Atlantic Canadian firms processed almost 200,000 metric tons of groundfish, over twice the output of New England groundfish processors in that year (DFO, *Annual Statistical Review*, 1988; USDOC, *Fisheries of the U.S.*, 1988). While almost all of the U.S. harvest is consumed domestically, most Canadian groundfish are exported. Over 84 percent of Canadian Atlantic groundfish landings and an even larger share of processed groundfish were exported in 1986, with over 82 percent sent to the United States (DFO, *Annual Statistical Review*, 1986).

Geographic Concentration

The size and location of Canadian groundfish stocks make the industry more dispersed geographically than the New England industry. Historically, large processing centers have been spread along the coast from Quebec and New Brunswick through western and southwest Nova Scotia and around much of Newfoundland. The restructuring of the corporate sector in the 1980s, followed by the current stock collapse, however, have led to many plant closings and increased geographic concentration.

Industrial Concentration

The industrial concentration of Canadian fish processing has been higher for many years than that of the United States processing industry, and it has increased in recent years. In 1968 the six largest companies accounted for a little over 50 percent of total groundfish sales, including sales of groundfish for some of the smaller processing companies (Kirby 1982). By 1980 the top four firms accounted for 63 percent of the groundfish processed and almost 70 percent of the groundfish marketed as smaller companies came to rely on the larger ones to market their products. The industry was even more concentrated in the frozen sector of the market, the top four firms marketing 90 percent of frozen fillets and 85 percent of frozen blocks.[8]

Concentration was further increased by government action in the early 1980s. The growth in landings and capacity in the late 1970s, following the establishment of the 200-mile limit, led to the rapid expansion of frozen-block output. Large processing firms relied on debt financing to expand much of their processing capacity. With the global recession of the early 1980s and an exchange-rate-induced shift in U.S. demand for frozen products from Canada to Iceland and Norway, demand fell and large processors became burdened by growing inventories of frozen blocks. These inventories were particularly expensive to maintain because of high interest rates (Kirby 1982; Barrett 1992b).

In the face of impending financial collapse, the processing companies turned to the Canadian government for help. C$15 million dollars were granted to the five largest companies on the condition that they merge into two corporations whose combined operation would account for about 75 percent of frozen groundfish production in 1985 (USITC 1986). An additional US$142 million in federal funds was used to purchase a 20 percent interest (plus $8.1 million in preferred stock) in the Nova Scotia company (National Sea Products) and a 60 percent interest (and $61 million worth of preferred stock) in Fisheries Products International in Newfoundland (USITC 1984). Smaller equity purchases were made by the provincial governments.[9]

The sharp declines in Canadian landings have again begun to put pressure on the large processing companies. Both have downsized their operations, closing all but one or two plants, and operating levels in the remaining plants have been substantially reduced (DFO 1993c). National Sea reported a loss of almost $36 million in 1991 and net income was down by almost 50 percent in the first quarter of 1992 (Taylor 1992). FPI reported smaller losses of $300,000 in 1991 and a loss of $1.5 million in the first quarter of 1992.

SCALE ECONOMIES, VERTICAL INTEGRATION,
AND INDUSTRY CONVERGENCE

Economic theory predicts that competition will lead to convergence around the most efficient production institutions. In whole-fresh-groundfish markets, this convergence appears to be happening.

Part of the Canadian fresh-fish wholesaling and processing sector, concentrated almost exclusively in southwest Nova Scotia, is using its proximity to the United States to market whole fish in New England markets, a product niche that has traditionally been neglected by the large Canadian processing firms. These processors and wholesalers are similar in scale and structure to those of New England. They are small (80 percent employ fewer than 100 workers) and have very low capital requirements (Apostle and Barrett 1992b); most have no freezing capacity (although they often have the ability to salt fish); and they are supplied with fresh-fish of high quality by a fleet of independent harvesters.

Divergent Institutions

At the same time, there is some indication that Canadian large-scale, vertically integrated processing firms are beginning to penetrate the U.S. market for fresh-fish fillets. This proliferation of competing forms of industrial organization in the fresh-fish sector raises anew the question of what the efficient scale and structure for the industry are.

The efficiency of the large corporate sector in Canada is based in part on economies of scale in the frozen-fish sector. Freezing capacity is capital intensive, effective marketing requires access to a large and steady supply of raw material, and the substantial concentration among the buyers of Canadian frozen-fish products in the United States market encourages countervailing concentration among producers. Moreover, Canada's major competitors in the international marketplace are also large and concentrated (Ministry of External Affairs 1983).

In Canada, restructuring and other public policies have favored vertical integration and increased concentration as a means of improving efficiency in frozen-fish processing. In fact, the primary justification given by the Canadian government for carrying out industry restructuring in the 1980s was that fewer large competing domestic companies would allow for scale economies and the more efficient production planning and marketing needed to maintain Canadian competitiveness in the international frozen-fish market (Kirby 1982).[10]

Large processors in Nova Scotia began to assert their presence in the fresh-fish sector in the mid 1980s, however, and National Sea Products is now the largest single supplier of fresh-fish to the U.S. market. These processors tend to export higher-valued fresh fillets, as opposed to the unprocessed whole fish of the much smaller southwest Nova Scotia wholesalers and processors. Fresh-fillet sales place the vertically integrated Canadian processors in direct competition with both the small southwest Nova Scotia companies and the New England processors and wholesalers for supplying large U.S. wholesale and retail markets.

One key to the Canadian processors' success in the fresh-fillet market is that they can command sufficient supplies to enable them to sell directly to large retail buyers in the United States. Large-scale buyers in the market prefer a steady supply of high-quality fresh-fish for marketing purposes, which presents a problem for smaller processing firms because they cannot secure a sufficient regular supply to allow them to compete. In fact, in a recent survey of supermarkets, 80 percent of the retailers listed inconsistency of supply as their major problem with the current small-scale processing-firm supply network (Hasselback and Marris 1991).

A second competitive advantage is that the large processors can eliminate New England brokers and processors as middlemen. The elimination of 'middlemen' has further increased the profitability of large-scale, fresh-fish processing over frozen processing.

Finally, U.S. tariff policy has favored large over small fresh-fish processors. For example, the countervailing duty imposed on fresh whole groundfish from Canada in the mid-1980s has had a disproportionate impact upon the small Canadian wholesalers and processors in southwestern Nova Scotia, because a similar duty was not imposed on the fresh fillets produced predominately by the large processors.

Will There Be Convergence around Large-Scale Production?

The industries in Atlantic Canada and New England evolved to serve two distinct markets – frozen and fresh. One consequence of this market segmentation was that industrial structure developed along different lines in the two countries. As the Canadian industry begins to penetrate the fresh-fish markets previously reserved for New England, the opportunity exits for competition to test the efficiency of these alternative production systems.

Both countries have similar atomistic industrial structures for serving the white-tablecloth and New England supermarket segments of the fresh market. In the potentially high-profit mass markets for fresh-groundfish fillets outside New England, however, the atomistic structure of the New England industry is in head-to-head competition with the vertically integrated and highly concentrated corporate sector in Atlantic Canada. A major question for the future is which structure will prove more competitive in this emerging mass-market segment.

Market Potential

There is strong evidence that U.S. mass markets for fresh fish have high potential for continued expansion. Fueled by a recognition of its dietary health benefits, seafood consumption per capita increased 22 percent between 1982 and 1989 (Food Marketing Institute 1990, 1991). This demand for seafood is being felt in the supermarkets, since 60 percent of shoppers in a recent survey listed the presence of a fresh-seafood section as important in determining their choice of a primary food store (Food Marketing Institute 1991). Fresh-fish sales in supermarkets have particularly benefited from this growth in seafood demand (Hasselback and Marris 1991).

In addition to the general expansion in demand for processed fresh fish, there is a potential for more rapid growth in U.S. markets located outside the New England region, especially in high-profit, mass-market outlets represented by supermarket chains in the midwest. Opening these markets will require a large and reliable supply of high-quality fish, however, a goal that has eluded the production systems of both New England and Atlantic Canada.

Currently, supermarket sales of fresh groundfish are largely confined to the northeast, where customers are familiar with a wide range of species and are accustomed to fluctuations in availability and price among different species. In such sophisticated markets, customers accept 'catch of the day' substitutions among species and, as price levels have risen, have come to accept price differentials over competing sources of protein such as chicken and hamburger. Building mass markets for groundfish in non-coastal regions, however, depends upon an initial ability to compete with meat in terms of price and to have sufficiently predictable supplies and prices to support large and repeated advertising campaigns.

Competing Forms of Industrial Organization

Given the potential growth in mass markets for fresh fish, which form of production is best suited to serve this market? The same structural characteristics that have yielded economies of scale for large processors in the production and marketing of frozen fish are starting to be a factor in the fresh-fish market in both Canada and New England. The strengthening of the large-scale corporate sector, however, has been heavily criticized by those who believe that large-scale harvesting and processing is inefficient (Apostle and Barrett 1992a).

It is argued that these large firms suffer from high fixed costs and bureaucratic inefficiencies, which force them to be constantly seeking public subsidies in order to stay afloat. In contrast, the smaller processing firms have knowledge of the fishing industry and the flexibility to adjust to the frequent fluctuations in species and landing volumes.

Thus, it is not clear that a massive increase in concentration in the fresh-fish industry will necessarily improve economic performance in the industry. Much depends on whether the gains in production and marketing efficiency can outweigh the costs of reduced flexibility and increased administrative overhead.

Although similar efficiency tradeoffs are also relevant in the United States, concentration in processing has remained low and there has been little formal vertical integration. New England processors and distributors are able to pool the erratic supplies of fish from the atomistic harvesting sector to smooth somewhat the day-to-day fluctuations in the supply of fresh fish. They have also turned to independent Canadian sources to meet their additional need for raw inputs. Especially during the winter months, Canadian whole fish account for as much as 70 percent of the total fish supply received by Boston distributors and processors and they are important to New Bedford processors as well. The pooling efficiencies available through New England processors and distributors, however, are insufficient to compensate fully for daily and seasonal fluctuations in landings, especially given the recent sharp decline in U.S. and Canadian groundfish supplies.

Instability in supply, therefore, remains a barrier to the full-fledged exploitation of mass markets for fresh fish outside the northeast. Neither the New England harvest, as augmented by that of independent vessels in southwest Nova Scotia, nor the Canadian corporate sector is able to provide sufficient supplies of high-quality fish at predictable prices to open mass markets in the midwest.

Nevertheless, several changes in the industry favor a possible restructuring of the processing market, with increasing scale and concentration the likely result. They include a reduction in transportation costs through improved air freight opportunities; a general expansion in demand for fresh fish in the United States in response to dietary health

concerns; and new forms of packaging and climate-controlled processing, which are extending the shelf life of fresh fish.

If New England stocks were larger, landings would be greater and
somewhat more predictable, so that the largest processors might be able
to expand to serve this market. At the same time, smaller processors
could continue to cater to local restaurant markets and to the many
small retail outlets in the Northeast that demand small quantities of top-
quality fish. On balance, larger, more stable fish supplies would likely
lead to development of an industry structure in New England that
would more closely resemble that of Atlantic Canada.

Alternatively, if the Canadian stocks were rebuilt, large Canadian processors might be able to exploit more fully the efficiency potential of the
enterprise allocation system in the harvesting of fresh fish. The corporate sector in Canada has the same capacity to exploit catch-pooling efficiencies as do the larger distributors and processors in New England.
Enterprise allocations permit intertemporal pooling, however, so that
the Canadian corporate sector has a greater ability to absorb output and
price fluctuations than is possible under the more loosely regulated
New England production system. Pooling both among vessels and over
time would allow the Canadian corporate sector to build a stable supply
channel of high-quality fish to serve mass markets throughout the
United States.

SUMMARY

While the fishing industries in the two countries harvest similar species
using similar technologies, in the aggregate their industry structures are
quite different. The Canadian industry is much larger than its U.S. counterpart and is a major participant in the world markets for frozen and
salted fish. This difference in size is largely a reflection of the difference
in the fishery resource endowments of the two countries.

Because of the relatively small size of the U.S. and Canadian fresh-fish
markets, the Canadian industry has relied primarily on frozen and
salted fish markets as an outlet for its products. The need to compete
with other large exporters of frozen- and salt-fish products has influenced the development of a highly concentrated fish-processing industry in Canada. As long as fisheries stocks are maintained at efficient
levels, the scale and scope economies achieved by large firms are sufficient to ensure participation of the industry in world markets.

Although the Canadian frozen-fish processing sector is characterized by vertical integration with harvesting and high concentration, a large inshore fleet made up of independently owned vessels is the major source of fresh-fish. This fleet sells to many medium-sized processors as well as to the large frozen processors. This sector is quite similar to that of the U.S. industry, which also consists of a large number of individually owned vessels selling to a large number of small processing firms. Thus, the fresh-fish processing sectors in both countries have exhibited similar industrial structures.

These similar structures have not yielded similar market arrangements for determining ex-vessel prices. Whereas the U.S. prices seem to reflect daily fluctuations in supply and demand through the use of auction markets and reference prices, Canadian fresh-fish prices are set mainly by brokers and dealers operating in the U.S. market or by the large frozen-fish processing firms that purchase fish from the inshore fleet. Prices to the large offshore fleet in Canada are set internally, since this fleet is owned by the processors.

Changes in the fresh-fish market in the mid-1980s resulted in the expansion of Canadian fresh-fish production and exports to the U.S. market. The current size of Canadian processing firms gives them an advantage over the much smaller U.S. firms in a potential growth market of large fresh-fish buyers because they can guarantee large supplies of fish. The U.S. firms are better able to ensure a uniform quality of fish, however, because their trips are shorter and they have been supplying this market for many years. Canadian firms need to expand their ability to deliver a consistent-quality product, since the quality required in the fresh-fish market is much higher than that needed in the frozen- or salt-fish markets.

A major factor in each country's ability to supply the U.S. fresh-fish market will be its success at managing its resources in the coming years. Both countries currently are suffering from overexploitation of their fisheries stocks, a handicap that is constraining processors' abilities to expand their sales.

4

Regulation and Industrial Policy

As fisheries managers in both countries face the immediate problem of restoring depleted groundfish stocks, they must do so in the context of the evolution of markets for fishery products, as described in the preceding chapter. The size of its stocks, its regulatory system, the structure of its industry, and its proximity to the United States provide the Canadian industry, in principle, with a competitive advantage in supplying a significant share of the United States fresh-groundfish markets. In particular, Canada can potentially provide large and regular supplies of fresh groundfish to emerging high-value markets in large retail outlets, while the U.S. regulatory system appears to be hindering the evolution of the fishing industry structure to compete better for these mass retail markets.

The success of the processing and harvesting industries in both countries will depend critically on the capability of U.S. and Canadian fisheries management policy to restore groundfish stocks in the coming years. Government regulation of the fishing industry in both countries is guided, at least in part, by the neoclassical economic theory of common-property resources. There is a large and complex technical literature on the application of this theory to the fishing industry (see Anderson 1986 for a textbook discussion and excellent summary and reference list of this literature), the basic elements of which provide a useful ground for comparing regulatory practices in the United States and Canada.

This chapter begins with a brief review of the economics of fisheries regulation (a more detailed discussion of the theory and its applications to policy is provided in the appendix to the chapter). The regulatory choices and other policy interventions made by the United States and

Canada are then examined. The chapter concludes with a discussion of the current problems and policy options facing fishery managers in the two countries.

EFFICIENCY THEORY OF FISHERIES REGULATION

The intuition behind the neoclassical economic theory of the fishery is that overfishing will occur whenever harvesting levels are unrestricted and there is relatively open access to the fisheries resource. When there are no restrictions on entry or harvesting, no single person or firm has ownership rights or control over these resources. Without individual ownership, there is no market incentive to husband stocks. Each fisherman has an incentive to harvest as many fish as is profitable, as quickly as possible, in order to beat competing fishermen to the resource.

The result is that too many resources are devoted to the fishing industry and the stocks are overharvested. In the extreme case, where the fishery is of particularly high value or where harvesting costs are low, so much effort may be devoted to harvesting that the biomass falls below the level required to reproduce itself and the species are threatened with extinction.

In more technical terms, the economic problem of overfishing can be measured by the extent to which the actual level of fishing effort (measured by the cost of labor, vessels and equipment, fuel, etc.) exceeds the level of effort required to reach the socially efficient 'maximum economic yield' (MEY) of a particular fish stock. The MEY level of effort equalizes the incremental (marginal) value of the harvest to society and the incremental cost to society of the fishing effort required to harvest the additional catch. MEY represents the socially efficient level of fishing effort because it provides the highest possible return to society per dollar spent on the labor, capital, and other productive inputs used in the fishery. Consequently, it yields the maximum economic rent (analagous to society's maximizing its profits from the fishery) – the difference between the value of the harvest and the cost of resources devoted to harvesting – to the economy.

In the absence of effective regulation or ownership rights, individual fishermen strive to obtain as much of this economic rent for themselves as possible by expanding their harvesting efforts. As harvesting effort increases, the costs of harvesting rise and the increased landings may drive down fish prices. The result is that the economic rent to society is gradually competed away, too many fish are harvested, and too many resources that could be used more productively elsewhere in the economy are devoted to the fishing industry.

Economic versus Social Management

If this scenario is to be avoided, according to neoclassical economic theory there must be management of the fishery that seeks to rationalize the industry by limiting the amount of resources devoted to harvesting the stock, usually by creating the equivalent of private property rights over the resource. Such management of the resource involves choosing the proper age (size) at which fish should be caught, the optimum amount of catch, the proper allocation of the catch over time, and maintenance of incentives to harvest the fish at the lowest possible cost (Anderson 1986).

Alternatively, some analysts argue that rationalization should occur through social processes, rather than by a centralized management process that creates some form of private property rights. They point to a body of case study evidence from around the world showing the effectiveness of community-based fisheries management systems, sometimes referred to as 'customary marine tenure.' These systems rely on traditional mores and social customs to govern access to, and use of, common-property resources without explicit regulation (Ruddle Hviding, and Johannes 1992; Santopietro and Shabman 1992; Quiggin 1988; Swaney 1990; Pinkerton 1989; Feeny 1990; Matthews 1993).

Social regulation has proved to be remarkably resilient against outside pressures to expand fisheries production and in many cases has avoided the overexploitation of fish stocks (Ruddle, Hviding, and Johannes 1992). Because a significant share of U.S. and Canadian groundfishing stocks are caught on offshore grounds and by vessels that are not amenable to such community-based controls, however, fisheries managers in both countries continue to opt for regulatory principles based, at least partially, on neoclassical economic theory (Matthews 1993; Edwards and Murawski 1993).

Distributional Implications of Fisheries Regulation

Rationalization of a fishery restores economic rents or 'profits' to society and transfers capital and labor from the fishing industry to other parts of the economy. Although moving from a situation of overfishing to a more rationalized industry can increase rents for those fishermen who remain in the industry, it also creates economic 'losers' within the industry as some fishing vessels go out of business and some fishermen become displaced. Determining which segments of the industry (i.e., inshore or offshore, recreational or commercial, processors or fishermen) should bear the brunt of reduction in harvesting required to achieve MEY, and whether the industry or government should receive the economic rent politicizes the regulatory process, creating incentives for undermining the more restrictive regulations.

Conflict over distributional outcomes often makes it politically difficult for government to implement fisheries regulations that are based solely on economic efficiency goals.[1] To date, neither the Canadian nor the United States government has authorized managers to focus solely on achieving economically efficient utilization of the resource. As a result, somewhat ambiguous goals have been developed – often an imprecise mix of biological, economic efficiency, and distributional criteria – in the legislation and regulations governing fisheries management in both the United States and Canada. This problem arises particularly under management systems (like that in the United States) that formally involve members of the industry in managing the fishery. Such systems often become gridlocked by interests competing for alternative distributional outcomes. Even in Canada, where the centralized management system is more insulated from special-interest politics, concerns for the economic survival of isolated fishing economies and with the allocation of revenue between the inshore and the offshore fleets are constantly modifying strictly biological or economic goals of fisheries management (Crowley and Palsson 1992; Halliday, Peacock, and Burke 1992).

REGULATORY OPTIONS

Rationalization objectives can be implemented through a wide variety of regulatory options. They include full privatization of the fishery (aquaculture); measures designed to increase the cost of fishing (open-access restrictions); setting aggregate catch quotas; limiting entry into the fishery; and the use of economic incentives, such as taxes and transferable quotas. As the following examples illustrate, each option has different implications in terms of efficiency and distribution.[2]

Aquaculture

Privatizing the ownership of fishery resources through aquaculture effectively negates common-property issues. Aquaculture is used in the

United States for oysters and clams and in Canada for salmon and at times for lobsters. Because of the wide-ranging migratory patterns and open-ocean feeding patterns of groundfish, aquaculture has rarely been applied to groundfish. (There are proposals, however, to develop large ocean fish pens that could be used to farm some groundfish species.) In terms of distribution, privatizing ownership through aquaculture initially gives the rent earned from the fishery to the owners of the aquaculture venture, but this procedure can be modified by licensing or other arrangements that would divide the rent between entrepreneurs and the government.

Open-Access Restrictions

A more common set of tools used by fisheries managers restricts access to the resource by limiting particular fishing technologies and harvesting seasons. These restrictions include the closing of areas of the ocean to fishing permanently or seasonally, controls on catching devices (such as requiring large-mesh nets), and prohibiting fishing for a particular species for specified periods of time.

Open-access restrictions reduce fishing effort by increasing the cost of fishing per unit harvested. While reducing fishing effort, they also reduce overall efficiency because they raise the cost to society of catching the existing fish stock. The increased cost of this effort means that such restrictions do not move the industry any closer to the MEY level of effort.[3] Furthermore, open-access restrictions do not limit the number of vessels or fishermen, and they encourage fishermen to devise even more inefficient, unregulated harvesting techniques, which can be substituted for those that are constrained by the management regulations.

The distributional effects of these restrictions are relatively neutral, or at least hard to predict, because all fishermen in a particular industry are subject to the same restrictions. While some fishermen will find it easier to adapt to such restrictions, it is not pre-determined from the outset who among them will be winners or losers. This perceived distributional neutrality may help to explain their popularity among fishing groups and thus their frequent adoption by government regulators.

Aggregate-Catch Quotas

Establishment of aggregate-catch quotas requires the setting of a total allowable annual catch (TAC) of a particular species and then the prohibition of further harvesting of that species for the rest of the year once the quota is filled. This mode of regulation is very inefficient, since it encourages fishermen to expand their catching capacity in order to harvest as much of the aggregate quota as quickly as possible. The cost of catching a given fish stock is thereby increased, and, once the quota is caught, there is a substantial waste of resources in the industry in the form of idle vessels and unemployed labor.

Like open-access restrictions, aggregate quotas are distributionally neutral. All fishermen initially perceive themselves as having an equal chance to harvest a substantial portion of the quota, and thus historically there have been minimal objections to the implementation of such a system. However, a quota eventually benefits those fishermen who can most rapidly increase the catching capacity of their vessels. Fishermen who fail to earn what they consider to be sufficient incomes before the closing of the fishery will likely put considerable political pressure on managers to relax the quotas as they become binding.

Limited-Entry Regulation

Limited-entry regulations are intended to reduce fishing effort by limiting the number of vessels or fishermen allowed into the industry. In theory, such regulations seem to be an easy answer to the common-property problem, but their application has proved quite difficult and can sometimes exacerbate the overfishing problem.

The primary difficulty is determining what dimension of effort to limit. If vessels are limited, fishermen have an incentive to resort to 'capital stuffing' or enlarging the fishing power of the vessels, since neither the fishing technology nor other inputs are limited. Controlling the quantity of fishing labor will lead to the same problem, because vessel owners will substitute unrestricted equipment inputs for restricted labor inputs.

A related issue is deciding how excess capacity should be reduced under the limited-entry regulations. For example, forcing vessels with lower historical catch levels out of the industry may eliminate part-time fishermen, who could be the most efficient harvesters in some fisheries.

Limited entry also raises several distributional issues, such as deciding which fishermen are allowed to remain in the industry, who will be allowed to enter, and (if limited entry is successful) how the economic rent generated by the industry will be distributed within the industry and between the industry and the government.

Economic Incentives

The regulatory options most preferred by economists involve the use of economic incentives – either a tax or *individual transferable* quotas. In theory, the tax should be placed directly on fishing effort. The practical difficulty of administering such a tax (including defining effort), however, means that it would have to be approximated by a harvesting tax on the value of the catch. If a harvesting tax is implemented, the least efficient fishermen will leave the industry, and, through adjustment in the tax rate, harvesting effort can be diminished until the MEY level of effort is reached. While such a tax reduces harvesting effort by increasing the fishermen's nominal harvesting costs (in contrast to the case where open-access techniques are used), it does not increase the social cost of catching fish, since no real resources are consumed in the process.

Although a harvesting tax is a theoretically effective means of rationalizing the harvesting sector, it has not been used, for both administrative and political reasons. The lags involved in collecting fisheries data and the uncertainties in interpreting such data make it very difficult to determine the tax level that will allow MEY to be reached. In addition, the tax transfers the economic rent from effort limitation to the government, so that fishermen remaining in the industry are no better off than they were before the tax was implemented. This fact combined with the inevitability of many fishermen being forced to leave the industry and the negative connotation the concept of 'taxing' carries with it make the tax politically very difficult to implement.

A second regulatory option that uses economic incentives is the individual transferable quota (ITQ). Shares of an aggregate catch quota are allocated among fishermen and they are allowed to trade these quota shares among themselves. Because individual quotas give each fisherman (or vessel) a right to a specific amount of the total fishing stock, ITQs also provide a way to privatize fishing stocks.

The advantages of this system are that the total quantity of individual quotas defines the maximum amount of fishing effort, while the transferability feature provides a market incentive for efficiency. If their quotas are transferable, those fishermen with lower fishing costs will have an incentive to buy additional quotas from those with higher fishing costs. The transfer of quotas from high-cost to low-cost vessels means that the total quota will be harvested at the lowest possible cost. Moreover, fishermen have incentives to invest in more efficient technologies, since they make more profits the less they spend on catching their quota.

ITQs have the same economic efficiency properties as the harvesting tax, but fewer of its liabilities. If the quotas are properly enforced, managers can be certain of the maximum quantity of the stock that will be harvested under the quota system, since this figure is determined by the size of the ITQs issued.[4] Moreover, ITQs can accommodate a wider variety of distributional outcomes than the harvesting tax. For example, auctioning the quotas will result in the government's receiving the rent from the fishery. In contrast, giving quotas to individual vessels will leave the rent with the fishermen, and under a transferable quota system the least efficient fishermen will sell their quotas to the highest bidders and leave the industry. In both cases, only the most efficient fishermen will remain in the industry over the longer term. The inevitability of the industry's downsizing, however, and the increased difficulty of entry into the industry (a quota would now have to be purchased from a current fisherman) make the initial implementation of such a system politically difficult.

IMPLEMENTATION OF FISHERIES REGULATION IN THE UNITED STATES AND CANADA

When distributional goals are combined with efficiency goals, it is apparent that moving to an 'optimal' fishery is a complicated political, as well as economic, process. This complexity is well illustrated by the way in which the theory of efficient regulation has been translated into regulatory practice in the United States and Canada.

The Canadian and the U.S. regulatory policies governing their respective North Atlantic fish stocks evolved from a common management regime, and both countries seemed to adopt implicitly the goal of moving toward the MEY target (Bannister 1989; Charles 1992; Matthews 1993; Edwards and Murawski 1993). The Canadians, however, have developed a centralized management structure that has actively pursued a combination of economic incentives to reach efficiency goals and direct intervention in industry structure to achieve distributional goals. In contrast, the United States has developed a more decentralized regulatory structure with minimal attempts to influence industry structure. The U.S. approach also has exposed the management process much more to distributional conflicts than has the Canadian, and this method has interfered with management policies aimed at rationalizing the industry.

Common Origins

Prior to the 1970s regulatory efforts consisted mainly of scientific research (coordinated by ICNAF) directed at better understanding the growth and behavior of fish stocks. Rather than trying to achieve MEY, limited management efforts were directed at loosely controlling the fishing effort of the participant countries in order to achieve a less efficient level of harvesting known as maximum sustainable yield (MSY) – the largest level of annual harvest that the fish stock can sustain without declining (see the appendix to this chapter for a detailed discussion comparing MSY with MEY).

As the adverse effect of increased fishing pressure became apparent during the 1960s, ICNAF members adopted open-access management techniques (Bannister 1989). Between the difficulty of enforcement and the fishermen's ability to switch to species for which less stringent regulations were in force, these regulations were not effective in limiting either effort or new entry into the industry.

In the early 1970s ICNAF established total allowable catch (TAC) targets designed to achieve MSY, which were apportioned annually among the member nations. However, there were no mechanisms to ensure that countries would abide by their TAC allowances. Even if a nation's vessels were found in violation of a regulation, there were no sanctions beyond notification of the appropriate government.

While Canada and the United States attempted to monitor their own vessels quite closely and fines were regularly levied for violations, several of the other nations were more lax in their enforcement. The result was considerable overfishing and eventually the dissolution of ICNAF. In its place, the United States and Canada each developed separate and distinctive national regulatory systems.

REGULATING THE CANADIAN GROUNDFISH STOCKS

Responsibility for Canadian fisheries management policy became centralized at the federal level and the minister of fisheries and oceans is given 'absolute discretion' to issue and authorize licenses for fishing by the Fisheries Act (Meaney 1992). The Department of Fisheries and Oceans (DFO) carries out the day-to-day management of the fisheries. The provinces continue to play a secondary role in fisheries management policy through their authority to regulate the processing industry and to initiate economic development activities that support the fishing industry.[5]

Setting Regulatory Policy

The Canadian approach to fisheries management is firmly rooted in biological analysis and the application of economic efficiency criteria. In 1976, Canada adopted the management objective of 'optimum sustainable yield' (OSY) (Environment Canada 1976). This objective has never been precisely defined, but it appears to encompass MEY efficiency criteria, which are then modified by distributional considerations. For example, the Kirby Task Force described the primary goal for OSY management as the economic viability of all sectors of the fishery, not as the maintenance of 'optimal' fish stocks (Kirby 1982). Other goals included maximizing regional employment in harvesting and processing and ensuring that fishery management plans contained elements of fairness in deference to the diversity of the Atlantic Canada fishery.[6] Thus, it appears that OSY is a management target that lies somewhere between MEY and MSY.

In practice, OSY is set in a multi-stage process. Initially a groundfish TAC is set for each groundfish species based upon the biological concept of MSY. The TAC is usually reduced to take account of the economic costs of harvesting, and then it may be adjusted still further (up or down) to meet the goals of employment opportunity and regional development (Copes 1982; Bannister 1989). In other words, regulatory targets are an ambiguous compromise among a number of goals – traditional biological goals for preserving the fish stock, economic efficiency criteria for reducing the level of resources devoted to fishing, and political pressures to provide a viable fishing industry in underdeveloped regions – which may conflict (Charles 1992).

Until 1993 basic management policy was set annually in the Atlantic Groundfish Management Plan, which took effect at the beginning of each calendar year. The plan specified the division of the TAC among processors, independent fishermen, and foreign fleets (Bannister 1989). Development of the annual management plan was a year-long process, which began with the preparation of initial TACs by DFO's Canadian Atlantic Fisheries Scientific Advisory Committee (CAFSAC). These TACs primarily reflected scientific data on the size of the stocks and their ability to survive various levels of fishing intensity, although the effort capacity of the existing fleets was also taken into account.

Normally groundfish TACs are set at a level referred to as '$F_{0.1}$.' This is defined as the harvest level at which the yield from the additional effort applied to the regulated stock would be at least 10 percent of that from an almost pristine stock subject to little fishing pressure (Atlantic Fisheries Service 1986).[7] The rationale behind $F_{0.1}$ is that any additional fishing effort beyond this point would result in a rapid decline in stocks. Prior to 1993 the $F_{0.1}$ TACs were incorporated into a draft management plan in which they were divided between inshore and offshore fishermen, based on the percentages each sector harvested in 1986.[8] This draft plan was submitted to the Atlantic Groundfish Advisory Committee (AGAC) (consisting of fishermen, processors, provincial government representatives, and DFO staff) for review and comment and was then sent back to DFO for possible modification.

With the depletion of the Canadian fish stocks, the process of developing an annual management plan was changed. In December 1992 the Fisheries Resource Conservation Council (FRCC) was established to replace the CAFSAC and AGAC. The council is appointed by the minister of fisheries and oceans and consists of representatives from all sectors of the fishing industry (offshore, inshore, processors, etc.), scientists from various disciplines (biology, oceanography, and economics), and ex officio members from DFO (FRCC 1993a).[9] The purpose of this restructuring was to make the process of developing recommendations on TACs and conservation measures to the minister more open to public scrutiny as well as to try to bring together all sectors of the industry at the beginning of the annual process along with the fishery scientists.[10]

Implementing Regulatory Policy

While Canada anchors its management plan on biological analyses and its overriding philosophy has been the preservation of stocks, Canadian managers have not been willing to implement economically efficient policies at the expense of the substantial displacement of the fisheries workforce. The result has been a patchwork of policies that combine those preferred by economists (such as ITQs) with less efficient forms of control, such as limited entry, seasonal closures, gear restrictions, and area closures.

The structure of the industry both hinders and enhances the application of these various techniques. For example, the concentration of processing activity among a few large firms permits an enterprise quota system that is much easier to enforce than a traditional transferable quota system applied to individual fishermen. Conversely, the presence of a large independent inshore fleet makes the imposition and enforcement of individual transferable quotas extremely difficult. The effect of these constraints on fisheries management is illustrated by the experience with the two primary measures used in Canada for limiting access to the fishery resource – limited entry and enterprise quotas.

Limited Entry

Canada first tried to control entry in 1973 by ending subsidies for the construction of vessels greater than 35 feet in length and by freezing licenses for offshore vessels greater than 65 feet in length (Bannister 1989). Rules were also put into place restricting replacement of existing vessels, limiting new entrants by requiring that they purchase an existing licensed vessel, and (in some cases) retiring the licenses of vessels that had left the industry.

Limiting entry into the offshore sector was relatively easy to enforce, because the vast majority of these vessels were owned by a few large processing companies. It was not as easy, however, to limit the development of new fishing technologies that increased the catching capacity of the offshore fleet.[11] As this increased capacity began to undermine the attempts to preserve fishery stocks during the early 1980s, a system of enterprise quotas (discussed below) was introduced to deal with the problem.

Beginning in 1977, DFO also began to confront the problem of limiting entry into the inshore fishery. Anticipating new resources becoming available, along with the 200-mile limit designation and the presence of unusually productive year classes of cod and haddock, new vessels entered the inshore fishery, which many claimed was already overcrowded. The number of fishermen increased by almost 50 percent and the number of vessels under 65 feet increased by 42 percent between 1977 and 1980 in Newfoundland and by over 6 percent between 1978 and 1980 in Nova Scotia (Copes 1982; DFO, *Annual Statistical Review*, various years).

In order to control this growth, licensing requirements became increasingly restrictive. Additional licensing of inshore vessels was frozen in 1980, and tighter restrictions on the size of replacement vessels were implemented (Hache 1989). The initial reaction of inshore fishermen to these restrictions was to replace existing vessels with larger ones. This move prompted fisheries managers to limit the length of new inshore vessels to under 65 feet. In response, fishermen increased the harvesting capacity of their vessels through more sophisticated fish-finding technology, expanded onboard fish storage capacity, and larger engines. The result was the building of a new fleet of high-capacity 'nearshore' vessels (60–65 feet in length) which were capable of fishing for longer periods and at greater distances than the traditional inshore vessels.[12]

When, despite large increases in groundfish prices, further restrictions in the mid-1980s made expansion of this size of vessel in the nearshore fleet less profitable, small boats (slightly under 45 feet) were replaced with 'jumbo' boats of roughly the same length, which had a much greater harvesting capacity (Hache 1989; Mandale and Morley 1990). To counter the effects of this expansion of the smaller-size vessels in the nearshore fleet, fisheries managers are now trying to regulate effort through restrictions on the cubic capacity of vessels (Hall 1990; Bannister 1989; Halliday, Peacock, and Burke 1992).[13]

These new nearshore vessels have the capacity to destabilize the industry. They can compete with the offshore sector for stocks of fish that were previously beyond their reach, which disrupts the delicate distributional balance among various inshore and offshore interests currently incorporated in the inshore/offshore division of TACs. Because of their superior technology, these vessels also threaten the viability of the traditional inshore fishery (Gardner 1988).[14] Older inshore vessels are finding their traditional stocks depleted by the nearshore fleet before the weather allows them to begin fishing.[15]

In summary, while the licensing program may be limiting the number of vessels in the Canadian fleet, it has not succeeded in reducing harvesting capacity and controlling overfishing. A DFO study of the Nova Scotia groundfish harvesting sector concluded that the fleet had the capacity to exert four times the effort required to harvest the TACs for the region (Bannister 1989; Halliday, Peacock, and Burke 1992; Burke et al. 1994).

Enterprise Quotas

The difficulties in controlling effort through TACs allocated to the inshore and offshore sectors led fisheries managers to begin an experimental enterprise quota allocation system for the offshore fleet in 1982.[16] Quotas were assigned to the individual processing companies with offshore vessels based on their past catch and projected ability to harvest resources. Following established principles of economic efficiency, the quotas were initially allowed to be fully transferable, so that the companies most efficient at catching a particular species could bid the quotas away from less efficient fishermen. Transferability was substantially restricted, however, after complaints that individual ports would be

arbitrarily and unpredictably affected by transfers of enterprise allocations among companies and fleets (Gardner 1988; Bannister 1989; Crowley and Palsson 1992).

Despite reports of cheating (through underreporting or unreported
sales by vessels while at sea), the quotas appear to have been a success.
The enterprise program has been extended twice since it was initiated
and is now scheduled to run through 1994. Moreover, it was extended
on a voluntary participation basis to the midshore fleet (65–100 feet) in
late 1988. Most importantly, the enterprise quotas encouraged the big
processing companies to spread out their catch throughout the year and
to respond more flexibly to changing demands for their products
(Crowley and Palsson 1992). The transformation that has taken place at
National Sea Products has been described as a shift in orientation 'from
a supply-driven to a demand-driven organization' allowing the company to focus much more on the quality of its fish products as opposed
to harvesting them as rapidly as possible (Gardner 1988).

The gains in efficiency from the effort restriction of enterprise quotas
have been reinforced by efficiency improvements in the marketing and
distribution end of the industry (Gardner 1988). Since the enterprise
quota system was confined to the offshore and midshore sectors, however, common-property inefficiencies remain in the fishery as a whole
and have been exacerbated by the development of the nearshore fleet
(high-capacity, 35–65-foot vessels), which threatens to impinge on enterprise quota allocations.

Current Management Problems

Despite the willingness to use a broad spectrum of regulatory measures,
including active attempts to restrict entry, Canadian stocks have still
suffered a substantial decline since the mid-1980s. From 1984 to 1990
TACs of all species shrank by an average of 24 percent and haddock
TACs in some areas were slashed by over 60 percent (DFO 1993b). In
1992 the northern cod fishery off the east coast of Newfoundland was
closed for two years, and this ban was extended indefinitely in 1994
(DFO 1994). Since 1993 most of the Atlantic groundfishery has been
closed and the TACs in the 1994 management plan have shrunk to 75
percent of their 1989 levels (DFO 1994).

The reasons for these stock declines are varied. One commonly cited
problem is that past TACS have overestimated potential catches by relying too heavily on reported catch data to assess the health of the stocks

(DFO 1993b). Reported catch data are often suspect, owing to misreporting and discarding of smaller fish or bycatch (species caught unintentionally while fishing for other species). In addition, fisheries scientists appear to have underestimated the increased catching efficiency of the vessels due to technological advances, so that high catches per unit of effort were incorrectly attributed to stock abundance when stocks were actually declining (FRCC 1993a; Safer 1994a, 1994b). As a result, TACs were set that allowed catch levels to exceed the 'true' $F_{0.1}$ level by two to three times (FRCC 1993a).

In the case of the northern cod fishery, much of the blame for the reduced cod population is being placed on overfishing by foreign fleets just outside Canada's 200-mile economic zone. Colder waters in this area and an increased seal population also are thought to have reduced the cod stocks (DFO 1992; FRCC 1993a; DFO 1993b).

In contrast, analysts have concluded that the chief problem with the Nova Scotia cod fishery is continued overcapacity in the harvesting sector, concentrated in the inshore and nearshore fleets (Hache 1989; Hall 1990; Halliday, Peacock, and Burke 1992).[17] In order to control this excess inshore capacity, the Nova Scotia inshore and nearshore fisheries now are divided into three sectors, each with its own regulatory system.

The largest sector consists of the traditional small-boat fleet, which fishes for a variety of species during the season, plus all inactive vessel licensees. The active vessels are subject to a maximum catch per trip in order to reduce the incentive to expand fishing capacity, and the number of trips are also limited. Inactive vessels are not allowed to transfer their licenses if they choose no longer to participate in the fishery.

The second sector consists of larger vessels (mainly those with fixed-gear technologies such as long-lining). The fixed-gear fleet is subject to an overall quota and would be closed down for the season once this quota was reached.[18]

The third sector includes the mobile-gear nearshore fleet (mostly 40–65 feet), which is responsible for much of the overcapacity problem (Burke et al. 1994). The task force recommended that this fleet choose among a variety of direct-effort limitation regulations (such as individual transferable quotas [ITQs] or an overall quota system) to regulate itself.[19] Beginning in January 1991, most of these vessels were brought under an ITQ regulatory scheme (Burke et al. 1994).

These policies maintain the basic regulatory structure that has already been established. They attempt to extend further the use of economic incentives, however, in order to reduce the effort and capacity of that portion of the fleet that has historically been most able to evade regulatory restrictions on catch.

REGULATING UNITED STATES OFFSHORE GROUNDFISH STOCKS

In sharp contrast to the situation in Canada, the U.S. federal government has historically played a small role in the management of U.S. fish stocks (Hennessey and Le Blanc 1985), and the current responsibility for fish stock management is shared among the federal government, state governments, and industry. Until the mid-1970s each coastal state had full responsibility for the management of the U.S. fish stocks in a zone that extended three miles from the coast. In response to U.S. fishing industry complaints about foreign fishing and the ineffectiveness of ICNAF, however, the Magnuson Fishery Conservation and Management Act was signed into law in 1976 and took effect in March 1977 (Anthony 1990).

The Magnuson Act established an Exclusive Economic Zone extending U.S. jurisdiction over marine resources to 200 miles from the shore. The states continue to have sole responsibility for managing the fish stocks within three miles of their coasts, while responsibility for the preparation of management plans for the balance of the fish stocks within the 200-mile economic zone rests with federal government and the fishery management councils in eight regions.

The composition of these regional councils allows for representation of diverse constituencies in the industry. The New England Fishery Management Council, for example, which has jurisdiction over the northwest Atlantic U.S. groundfish stocks, is made up of seventeen voting members – the NMFS regional director, the state fishery directors of each New England coastal state, eleven representatives of the harvesting and processing industry appointed by the U.S. secretary of commerce, and four non-voting members from federal agencies, such as the Coast Guard and U.S. Fish and Wildlife Service. The council has several committees, including a standing committee of industry experts who review proposed management measures.

Although the regional councils are responsible for preparing preliminary regulatory plans, NMFS retains authority over the final plan through its responsibility for reviewing all council plans, judging their scientific and technical merit, and advising the U.S. secretary of commerce as to their compliance with the national standards established by the Magnuson Act.[20]

Setting Regulatory Policy

The legislated standard for managing the fish stocks is known as 'optimum yield' (OY), the level of fish stocks that yields the greatest benefits in terms of production and recreation. OY is defined as the biological concept of MSY 'as modified by any relevant economic, social, or ecological factor' (Magnuson Act, as amended, 1980). Because it balances MSY against other considerations and is not a precisely defined target, OY resembles the OSY concept used in Canada.

The conflicting mix of objectives embodied in the OY concept is illustrated by the diversity of legislative guidelines for the development of management plans. These guidelines include the utilization of the best available scientific information as a basis for conservation and management measures; the promotion of efficiency in the use of fishery resources, *'except no such measure shall have economic allocation as its sole purpose'*; and the recognition of variations among fisheries stocks and catches. The restriction on economic allocation appears to be an explicit recognition of the possible conflict between efficiency and distributional goals of fishery management.

The development of a groundfish management plan to implement OY is a cumbersome and time-consuming process. It begins with biological data that are provided by NMFS. In the New England region, these data are reviewed by the council's professional staff and a five-person groundfish oversight subcommittee, who then develop the outline of the management plan, establish an agenda, and identify further research needs for the development of a full plan.

The full council reviews the plan at different stages of development from approval of the objectives to endorsement of the management techniques and regulations for achieving the objectives. After the council approves a 'draft' fisheries management plan, it is subject to public review and comment and may be revised before being forwarded to the NMFS for further review. In the past, if the Department of Commerce found that a plan did not meet the national standards specified in the Magnuson Act, the plan was rejected and returned to the council for amendments and another round of public hearings.

In contrast to the Canadian development of an annual plan, the entire U.S. process can take as long as two to three years. Since 1977 only two groundfish management plans have been implemented – one in 1982 and another in 1986. Although the 1986 plan has been amended five times, only Amendment #5 of March 1994 has significantly altered the management plan. The length of time required to develop and implement management plans makes the U.S. regulatory system much less flexible and much slower to respond to changes in fish stocks and fishing effort than the Canadian management framework.[21]

Implementing Regulatory Policy

While the Canadians have used the relatively efficient practices of limited entry and transferable quotas for managing groundfish, the New England Management Council (until Amendment #5) has relied upon less efficient approaches. They have included overall species quotas and effort restrictions that do not directly limit entry into the harvesting industry (Acheson 1984; Edwards and Murawski 1993; NEFMC 1993).

One of the main obstacles to adopting more economically efficient regulatory methods has been the ability of vested economic interests to be articulated in management policy through direct involvement of the industry in determining its regulation. The Canadian management system has been more shielded from these pressures by the lack of a direct role for the industry in developing the management plan.

The New England fishing industry has been strongly opposed to limited entry, even though its successful implementation would make those remaining in the industry better off. One reason for this opposition, voiced by many fishermen, is that entry restrictions undermine the independent nature of the harvesting industry which attracted them to the fishery in the first place. Some members of the council also have expressed concern that limited entry will restrict the opportunities for friends and relatives to enter the industry and will inevitably force some fishermen to leave it. Another argument, which has moved to the forefront in the early 1990s, is that the ecological system that controls fish stocks is too complicated, and our understanding of it is too rudimentary, to support limited entry or transferable quotas (Wilson 1992; *Commercial Fisheries News* 1992).

As an alternative to limited entry, the New England Management Council initially experimented with species quotas to limit fishing effort. By the fall of 1977 the binding quotas limited fishermen to small amounts of catch per trip, and when harvest levels continued to rise, the haddock, cod, and yellowtail flounder fisheries were closed (Dewar 1983, 1986).

Fishermen protested the closure locally and in Washington, DC, and lobbied successfully for an increase in the quotas. It is alleged that fishermen also developed tactics for evading the quotas, such as night landings, mislabeling landings, and misreporting catches (Anthony 1990). The result was the abandonment by the council of quotas as a regulatory tool in the early 1980s (Foreman 1982; NEFMC 1993; Anthony 1990).

In place of quotas, the council opted for 'open-access' regulatory policies, such as mesh-size regulations by species, seasonal closures for spawning, limitation of the size of vessels to be used in specific locations, restrictions on certain gear types in particular locations, and restrictions on by-catch and discards. While the intent of such regulations is to protect the fish stocks, their economic effect is to increase the cost of fishing by restricting the technology and location of fishing activity. Such increases in fishing costs might reduce the amount of fishing effort applied to the stocks, but in the case of the New England groundfishery they have motivated fishermen to devise new ways of circumventing the regulations, and thus fishing effort has increased (Massachusetts Task Force 1990; Anthony 1990).

Current U.S. Management Problems

The United States, like Canada, is going through a 'crisis,' which many fisheries policy-makers and managers now blame on overfishing. There have been large increases in days fished in New England since the passage of the 1976 Magnuson Act, owing to increases in the number of vessels; and there have been technological advancements that have substantially increased the catching efficiency of each hour of effort. As a consequence, fishing effort more than doubled between 1977 and 1987 (Massachusetts Task Force 1990; Anthony 1990).

Renewed calls for stricter regulatory policy resulted. One study, for example, recommends that managers adopt a goal of restoring the offshore groundfish stocks to their pre-1960 levels and encourages the U.S. secretary of commerce to intervene if the New England Management Council cannot achieve this goal in a timely fashion (Massachusetts Task Force 1990). It further urges the reinstatement of species-specific quotas combined with individual vessel trip limits; expansion of enforcement efforts and the revision of regulations to reduce the cost of enforcing them; expansion of efforts at assessing stocks and setting catch limits; and closer examination of the use of taxes, individual transferable quotas, and limited entry as instruments for reducing harvesting effort.

The urgency of these recommendations is underscored by the discovery of a possible major ecological shift on Georges Bank, where a large expansion of skates and dogfish is taking the place of the overharvested traditional groundfish species. In addition, scientists responsible for assessing the status of Georges Bank groundfish stocks issued a report in August 1994 indicating that the cod, haddock, and yellowtail flounder spawning stocks were at dangerously low levels (Plante 1994).[22]

In response to the 1990 findings, the Conservation Law Foundation and the Massachusetts Audubon Society filed a federal suit against the U.S. secretary of commerce on 28 June 1991 for failure to prevent overfishing of the New England groundfish stocks. This suit was settled with the proviso that if the New England Management Council failed to develop a final plan by September 1992 to restore cod and flounder populations to their pre-1960s levels within five years, and haddock populations within ten years, the secretary of commerce must implement such a plan by November 1992. Initial estimates by the council's groundfish committee indicate that a 50 percent reduction in fishing mortality is required to comply with these management goals (Plante 1992).

Even under the pressures of the consent decree, the problems created by having the industry manage itself through representation on the council became obvious. While federal officials were strongly advocating direct effort reduction measures, the industry representatives remained opposed to quotas and limited entry. Instead, they have proposed more stringent indirect measures to control fishing effort, such as further increases in the mesh size, increases in the minimum fish sizes, and possible additional area and seasonal closures (NEFMC 1993). This conflict in objectives and methods of reducing fishing effort prevents the council from instituting immediate emergency measures to restrict harvesting. In contrast, under the Canadian system TACs were immediately cut in response to claims of overfishing.

Although the council released its draft amendment to the management plan in March 1992, heavy political pressure from the industry resulted in a negotiated one-year extension of the deadline for the new management plan with the Conservation Law Foundation. The final draft of Amendment #5 was submitted by the NEFMC to NMFS in September 1993 with the goal of reducing fishing mortality on the major groundfish stocks by 50 percent within five to seven years. NMFS rejected parts of the plan because of its too generous limit on possession of haddock, and in early 1994 NMFS took an unusual emergency action and closed the Gulf of Maine and Georges Bank to haddock fishing. In May 1994 the NEFMC voted to limit haddock possession to the level recommended by NMFS and Amendment #5 was approved in full. Significant area closures on Georges Bank have been subsequently extended to cod stocks, and more stringent management measures are being considered.

Amendment #5 represents a substantial change in management policy, since the plan introduces limited entry to the New England groundfishery for the first time, by placing a moratorium on otter trawl vessel permits as of February 1991 in addition to establishing maximum trip limits for haddock. Moreover, the plan introduces effort reduction by requiring every vessel over 45 feet in length to reduce its number of days at sea by 10 percent each year for five years or to declare itself 'out' of the fishery for blocks of time. Each block is a minimum of twenty days, totalling eighty days the first year plus one additional day for every two days of fishing time, with increasing blocks of time in consecutive years, totalling 233 non-fishing days by 1999.[23] The new plan also requires larger mesh sizes and sets limitations on the upgrading of vessels.[24]

On the surface, Amendment #5 is economically inefficient, since, by its provisions, all vessels will be idle during some part of the year when normally they would be fishing. However, the plan is likely to force marginal vessels, for which 10 percent effort reduction threatens their economically viability, out of the industry. There has already been discussion within the council of allowing these marginal vessels to sell their fishing rights to more profitable vessels, and vessel buyout plans are also being considered. If the sale of fishing rights is allowed, the system would evolve into an individual transferable quota plan. In this case, the most inefficient vessels would leave the industry and the remaining vessels would be able to fish full time.

The process of fisheries management is also being examined at the federal level, since the Magnuson Act is up for reauthorization in 1994. For example, the Clinton administration has submitted a reauthorization bill that alters the definition of optimum yield (OY) to require a rebuilding of the stocks to MSY before modifications for social, economic, or ecological factors are allowed.[25]

ECONOMIC DEVELOPMENT POLICIES AND INDUSTRY SUBSIDIES

Management of the fisheries is not the only form of government intervention in the industry in the United States and Canada, and regulatory shortcomings have not been the only cause of overexpansion in the industry. The combination of underdevelopment in fishing communities and their dependence on fishing-related activities, has given the fishing industry a special prominence in economic development and industrial policy in both countries. Typically, these programs have encouraged fisheries expansion that runs counter to the goals of fisheries management.

Canadian Policies

The level of resources, the range of policy instruments, and the overall integration of development policies is far greater in Atlantic Canada than it is in New England. As of 1986 the U.S. International Trade Commission identified twenty separate programs that subsidized the Nova Scotia fishing industry, including those directed at regional development through Canada's Department of Regional Industrial Expansion (DRIE). Moreover, the restructuring of the corporate processing sector (described in chapter 3) involved massive infusions of government funds into the industry.

The greater intensity of economic development policy in Atlantic Canada must be understood primarily in terms of a broad national and provincial commitment to sustaining employment and income in isolated and slowly growing regions. For example, DRIE was created to 'increase overall industrial, commercial, and tourism activity in all parts of Canada and in the process, reduce economic disparity across Canada' (Ministry of State for Economic Development 1984).

This commitment preceded much of the regulatory apparatus of the fishing industry. Until the late 1980s DRIE provided financial assistance in the form of grants, loans, and loan guarantees for all aspects of the industry's 'life-cycle' – product development, new plants, plant modernization or expansion, and marketing. The eastern Canadian fishing industry benefited substantially from DRIE programs, since it was the primary creator of jobs in this economically disadvantaged region. Under the DRIE programs, assistance was available for the development of docks and other port facilities, such as cold storage and ice making, and for the modernization and construction of processing plants. These federal programs were supplemented by counterpart provincial programs.

Economic development subsidies have been further reinforced by national and provincial industrial policies. Until 1986, for example, the Fishing Vessel Assistance grant program (targeted at the small vessel fleet under 75 feet in length or around 100 tons displacement) provided funds for up to 25 percent of the cost of purchase or construction of vessels. In addition, vessels from 45 to 75 feet were eligible for 50 percent subsidies on the purchase of new equipment up to a maximum of C$850.

There are counterpart loan guarantees and grant programs at the provincial level, as well as subsidies for purchases of new fisheries equipment. With the exception of the loans associated with the restructuring of the large-scale processing sector, most of these subsidy programs are directed toward assisting the smaller, inshore vessels, and they are designed to aid the isolated port economies that are so dependent on the fishing industry. Examples include a low-interest loan program for fishing vessels in New Brunswick, a grant program for the construction of new inshore vessels in Newfoundland, and low-interest loans for fishing vessels and 50 percent cost-sharing industrial development grants for researching or acquiring new technology and constructing new processing facilities and vessels in Nova Scotia.

Government subsidies to the fish-processing industry also have been a long-standing practice in Canada. Between 1969 and the major restructuring in 1982, 260 establishments had received over C$46 million of Regional Development Incentives Act funds, covering almost 30 percent of the cost of their capital improvements. In addition, the federal government provided C$140 million in special assistance to processing firms between 1974 and 1977, and several provincial grant and loan programs were implemented during this period (Kirby 1982).

The recent problems of reduced landings and industry overcapacity have resulted in many of these programs being cut back or phased out. In their place, are adjustment and income support programs for fishermen and processing workers left unemployed by the closures of fishing grounds. Under a five-year, C$1.9 billion program called The Atlantic Groundfish Strategy (TAGS, initiated in May 1994), all fishermen and processing workers with 'substantial historical dependence' on the groundfish industry are eligible for weekly benefits ranging from C$200 to C$382. TAGS replaced two programs that had been in operation during the previous year – the Northern Cod Adjustment and Recovery Program (NCARP) and the Atlantic Groundfish Adjustment Program (AGAP). The program is also designed to include buyouts of fishermen's licenses and early retirement provisions, although as of the summer of 1994 the process for implementing these provisions had not yet been determined (*Commercial Fisheries News*, 1994).

U.S. Policies

The types of broad-gauge and coordinated fisheries development policies present in Canada have never existed in New England. The primary policies designed to aid the fishing industry are subsidies under the Federal Vessel Obligation Guarantee (FVOG) and the Capital Construction Fund (CCF) programs, both of which have been in existence since 1970 and are operated by NMFS. Local economic development initiatives involving the fishing industry have been limited largely to the improvement of port and harbor facilities, such as the $2 million port redevelopment project in Provincetown and the development of a display auction at the Portland fish pier.

Until 1986 FVOG provided loan guarantees for up to 87.5 percent of the cost of constructing or reconstructing any fishing vessel over 5 tons. This program was most actively used by the industry during the late 1970s, when many vessel owners were upgrading to more modern vessels. The number of approved loans increased from 120 in 1977 to 325 in 1979 and then declined rapidly (Corey and Dirlam 1982). Processors have also been eligible for loan guarantees under the program since 1982, but only a few have applied. Currently FVOG loan guarantees are restricted largely to vessels or processors committed to the harvesting and processing of underutilized species (U.S. Fish and Wildlife 1992).

The CCF program provides 'interest free' loans to fishermen through tax deferral on income deposited in a fund for the construction, reconstruction, or acquisition of a fishing vessel.[26] There were 131 New England vessels involved in the program in 1979, but this total had declined to eighty-four in 1994.[27]

Indirect federal assistance is provided by the National Oceanic and Atmospheric Administration through various marketing and technical advisory services, and the industry is also protected by the 'Nicholson Act' (46 U.S.C. 251), which prevents foreign fishing vessels from landing fish in U.S. ports. This law gives the New England fishermen an advantage in the regional market, since it delays the arrival of Canadian-caught fish by at least one day, but it is less important for more distant U.S. markets that receive their fresh-fish supplies by air.

As its Canadian counterpart did, the U.S. government responded to pleas from fishermen for aid in order to help them adjust to the stock collapse and the implementation of Amendment #5. In contrast to the provisions of the Canadian assistance plan, however, no direct income subsidies to fishermen are involved. Instead, the monies focus on revitalizing port communities and diversifying their economies.

The U.S. Fisheries Reinvestment Act of 1992 authorized appropriation of funds to fishing communities for pilot projects that focus on alternative fishing opportunities, aquaculture, and the development of markets for underutilized species. In 1994 US$1.5 million was appropriated for the five major New England ports, one port in New Jersey, and two ports on Long Island, New York. An additional US$1 million was authorized to help communities to develop proposals for pilot projects.[28] In addition, in 1994 US$30 million was made available to the New England industry from federal disaster assistance funds, although not in the form of individual subsidies. Sixty percent of these monies was earmarked for revolving loan funds and grants in communities most severely affected by the sharp decline of fish stocks, and the balance was used for loan guarantees (such as those made through the FVOG program) and direct grants to fishermen designed to stimulate investment in alternative economic activities.

SUMMARY

The United States and Canada have dramatically different approaches to the management of their fish stocks. Canada uses a highly centralized approach, with the federal government making most of the decisions regarding the size of yearly harvests allowed for different species and how the harvest will be distributed within the fishing fleet. The United States, in contrast, delegates many of the management decisions to regional management councils that include heavy representation from industry. In addition, there is an elaborate public hearing process required for any management decisions, whereas in Canada the hearing process is more of a courtesy than a legally binding process.

As a consequence of their differences in management structure, the two countries have developed a different set of regulatory tools for controlling harvesting effort. Canada has utilized several direct control techniques suggested in the economic literature, such as limited entry and privatization of the fish stocks through the use of individual transferable quotas. It has been able to apply the transferable quota system at the processor level for the offshore fishing fleet because of the vertically integrated structure of the offshore harvesting industry. Early evaluations of this system are quite positive.

Until quite recently, the New England Fisheries Management Council avoided direct controls on harvesting effort and focused instead on indirect techniques that raised fishermen's costs, such as area and seasonal closures and large mesh sizes. Continued overfishing recently forced the council to implement direct regulatory controls in the form of limited entry and restrictions on fishing effort (by limiting days at sea). The continued close involvement of the fishing industry in managing itself, however, will make the successful implementation of these controls exceedingly difficult.

Both Canada and the United States still face overfishing problems. Canada is trying to remedy these problems by expanding its direct control regulations to the entire fishing fleet, but it has also sharply reduced the TACs for various species in recent years. In trying to extend the transferable quota system to the inshore fleet, the lack of vertical integration has forced Canada to apply the quotas at the vessel level rather than the processor level. It is too early to determine the success of this regulation in reducing the fishing effort of the inshore fleet, particularly because most of the fishery has been closed since these regulations were implemented.

It is similarly difficult to predict the likely outcome of the current U.S. attempts to control overfishing. Successful attempts to limit harvesting effort will increase the economic rent earned in the fishery and thus the pressures to overfish. The current effort restrictions will have to be strictly enforced. Moreover, unless they evolve into some sort of transferable quota system, they will be quite inefficient, since substantial amounts of capital and labor dedicated to fishing will remain idle during much of the year.

Inadequate fisheries management policies are not the only source of the current overcapacity facing the fishing industry in both countries. Subsidy programs, particularly in Canada, are a further indication of the conflict between promoting economic development in port economies and achieving regulatory efficiency in the fishing industry. The Canadian vessel subsidies, during the late 1970s and early 1980s, and U.S. subsidies, to a more limited extent, contributed to the present-day excess capacity of the fleets in both countries.

APPENDIX

THE SIMPLE ANALYTICS OF FISHERIES REGULATION

This appendix is intended to give a brief summary of the key analytic elements of the economics of fishery regulation. For the sake of brevity, the discussion will be framed in static terms, but the critical differences introduced by a dynamic model will be noted. A more detailed textbook discussion of these concepts can be found in Anderson (1986).

Simplifying Assumptions

Several assumptions will simplify the discussion and highlight the important factors involved in regulating the fishery. First, it is assumed that changes in the market supply of fish resulting from a reduction in overfishing will be small enough to leave the price of fish unchanged. This assumption is appropriate for the New England stocks, since potential imports of fresh fish from Canada are large enough to keep any New England supply reduced through fishing effort restrictions from substantially increasing the market prices. Relaxing this assumption would shift out the average and marginal revenue curves for the industry so that the MEY point would be reached with less total effort reduction than if prices had remained constant (figure 4A.1) but does not otherwise affect the underlying analytics.

The second assumption is that the marginal cost of fishing effort (the additional labor, capital, and energy resources used to increase the harvest rate of the fish stocks) is constant. In other words, the cost of inputs do not change with the amount demanded and that the marginal and average costs of effort are always equal. The purpose of this assumption is to simplify the analysis by allowing the same curve to be used to represent marginal and average costs, but it does not alter the basic conclusions of the theory.

The final assumption is that the return to each additional unit of harvesting effort will diminish with increasing effort. Thus each additional unit of effort yields less fish than the previous unit of effort because of the smaller remaining stock of fish. The consequence of this assumption is that the marginal productivity of an additional unit of effort is always less than the average productivity of effort.

The Static Model

The three key concepts for understanding the economics of fisheries

Figure 4A.1
Regulating fishing effort

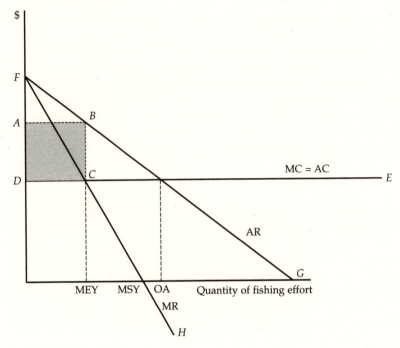

regulation are (1) maximum economic yield (MEY), (2) maximum sustainable yield (MSY), and (3) open-access equilibrium (OA). The levels of fishing effort associated with these three concepts are shown on the horizontal axis in figure 4A.1. The vertical axis measures revenues and costs in dollars.

The marginal/average cost curve is represented by the horizontal line *DE* (under the assumption that each additional unit of effort costs the same amount of society's resources). The marginal revenue per unit of effort curve (*FH*) is derived by multiplying the incremental catch resulting from an additional unit of effort by the price of fish. The average revenue curve (*FG*) is derived by dividing the total value of catch by the corresponding level of effort. Because of the assumption of diminishing returns to fishing effort, both the marginal revenue and the average revenue curves are negatively sloped.

From society's point of view, the optimum level of effort is reached at the intersection of the marginal cost and marginal revenue curves (*C*), where an additional expenditure of resources on catching fish yields an equivalent value of fish. This level of effort corresponds to MEY. At effort levels below (to the left of) MEY, additional effort yields a greater value of fish than their costs of harvesting, while at effort levels greater than (to the right of) MEY, additional harvesting or effort costs are greater than the resulting benefits in terms of the value of fish harvested.

Another way to interpret MEY is that it represents the level of effort where society obtains the maximum profit (or rent) from its fishery. In figure 4A.1, this rent is indicated by the shaded area, *ABCD*. In quantitative terms, the rent is the difference between the total revenue and total costs of a given level of fishing effort. Since no other level of fishing effort will yield greater rent to society, the important policy implication is that any regulatory action that moves a fishery toward the MEY level of effort will increase the rent earned in the industry.

As can be seen from figure 4A.1, the maximum sustainable yield (MSY) level of effort is where the maximum amount of fish can be harvested annually without reducing the population of the fish stock. The marginal revenue from additional fishing effort is zero at MSY, indicating that total revenue from the fishery is maximized. The costs of the additional harvesting effort required to reach this point from MEY, however, outweigh the social benefits of the increased catch. Although total revenue is maximized, the difference between total revenue and total costs is not maximized, since marginal costs are greater than marginal revenue. MSY will always be reached at a higher level of effort than that associated with MEY as long as harvesting costs are greater than zero.[29]

OA is the level of effort that arises in the absence of regulation. Without regulation (or the private ownership of the resource), no individual harvester is able to secure a share of the rent generated by an efficiently managed fishery and thus has no incentive to take this rent into account in deciding how much effort should be spent on harvesting. Since profits exist as long as total revenue from the fishery is greater than total costs (or, equivalently, as long as average revenue exceeds average costs), the prospect of gaining extra profit will cause the amount of effort devoted to the fishery to increase up to the point where all rents or profits are eliminated.

Because effort levels to the left of OA yield profits, and additional effort beyond OA results in losses, the open-access fishery gravitates toward an equilibrium at OA where no further increase in effort devoted to harvesting is profitable. While OA effort levels do not necessarily result in the extinction of the resource, from society's point of view far too many resources are being devoted to the fishery. (In the case shown in figure 4A.1, the marginal revenue of additional levels of effort is negative at OA, as it is at any point with higher effort levels than MSY.)

Dynamic Models

The model described here assumes a static fishery and shows the effort level that would generate maximum net benefit in any one period. As long as there is a positive discount rate, MEY is at a higher level of effort than the static model indicates, since people value fish consumption today higher than fish consumption in future years. Most studies have shown, however, that the dynamic MEY is still at a level of effort below MSY (Clark 1976).

In addition to taking into account the effect of the discount rate, proper resource management in a dynamic framework requires consideration of how the value and costs of harvesting are likely to change over time. For example, although effort levels of OA are not likely to result in extinction of the resource, since costs are sufficiently high to halt increases in fishing effort before the stocks are totally depleted, pressures on stocks can increase over time because OA effort levels will rise if the price of fish or productivity of effort increases or if harvesting costs fall.[30] Likewise, OA will occur at lower levels of effort if costs rise substantially.[31] Open-access regulatory controls, such as increased mesh size or seasonal closures, increase fishermen's costs, causing the OA level of effort to fall; but they do not move the fishery to the MEY level of effort, because rising costs also cause MEY to decline.[32]

Application of the Theory

During the 1950s and 1960s there was a lively debate between fishery biologists and economists about how to apply these regulatory concepts. The biologists advocated the higher, MSY, level of fishing effort, while the economists argued for the lesser, MEY, level of effort on the grounds that biologists were assuming harvesting could be undertaken at no cost. Much of this discussion remains academic, however, because of the imprecision in measuring and predicting changes in the levels of fish populations and because most fisheries have found themselves much closer to the open-access point, owing to the accommodation of distributional goals and the difficulties in enforcement.

In practice, several problems arise in managing a fishery with an MEY goal in mind. First, biological data on fish stocks are subject to long lags in reporting and analyses and are often inadequate for precisely determining the relationship between harvesting effort and stock levels. Thus, management decisions are often made with inadequate information on the health of the fish stocks.

Second, capital and labor inputs often do not move easily in and out of the fishery in response to changes in stock levels or management decisions. This factor 'stickiness' complicates the estimation of the resource costs of different levels of effort that are needed for the determination of MEY (Terkla, Doeringer, and Moss 1988).

Third, it is difficult to target fishing effort uniquely on specific stocks. Because by-catch of other species is inescapable, management changes directed at one species will often have major ramifications on alternative species.

Finally, management goals are not driven solely by efficiency considerations; they also incorporate the distributional consequences of management decisions. These consequences may be severe enough or politically unpalatable enough to make the MEY goal impractical. Determining which sectors of the industry will be contracted in order to move away from an overfishing situation is often so fraught with political conflict that proposed restrictions on effort are continually delayed.

Moreover, even if regulators do have success in restricting effort, rent will be created and individual fishermen will attempt to capture this rent through increased effort. Successful regulation also means that decisions must be made about how the rent generated by properly managed resources will be distributed between government and the industry and within the industry.

5

Workplace Employment Systems and Labor Market Structure

The cost of labor is an important factor in both the harvesting and the processing industries. Data from Canada's harvesting sector, for example, show that the crew's share absorbs between 17 percent and 26 percent of total revenue and, after the captain's share, is the largest component of harvesting cost (DFO, Scotia-Fundy Region, 1991). Processing also has a high level of labor intensity: wages and salaries account for 20–25 percent of total onshore production costs and about 57 percent of production costs net of materials (Kirby 1982; DFO 1993c, table 13-1).

The various components of labor costs in the fishing industry – wages, labor productivity, employment levels, and workforce attachment – are influenced by employment institutions as well as market forces. In this chapter the institutional structures of harvesting and processing labor markets are examined. The diversity of employment systems found in the fishing industry – kinship, paternalistic, and corporate – are described, and the efficiency and distributional consequences of these different systems are explored.

HUMAN RESOURCES PRACTICES IN HARVESTING

Workplace human resources practices in the harvesting sector vary widely according to the size and ownership of vessels. Large, corporate-owned vessels tend to adopt relatively formal practices, whereas family-owned and smaller vessels have informal employment systems that are often governed by kinship considerations.

Recruitment of Labor

Larger vessels tend to rely on formal recruitment channels, and crew are often selected on the basis of skill and formal qualifications rather than kinship (Andersen 1972a). Even on large vessels, however, family recruitment networks can be a part of the staffing process. The offshore Gloucester fleet, for example, is almost entirely dominated by first- and second-generation Italian families that captain their own boats and staff their crews with relatives and friends, and an increasing share of the groundfishing fleet in New Bedford is owned and operated by Portuguese families (Doeringer, Moss, and Terkla 1986b). Similarly, a study of offshore vessels in eastern Canada found that almost half (46 percent) of the captains and over one-third (36 percent) of the deckhands had relatives working on the same vessel (Proskie and Adams 1969).

On smaller vessels, families are often the major source of labor supply. Crews are recruited through kinship networks, augmented by friends and others from the community (Jorion 1982; Doeringer, Moss, and Terkla 1986b; Stiles 1979).

Job Content and Training

Differences in size and gear necessarily affect the specific content of work, but fishermen on all types of vessels work in teams with relatively little hierarchy or formalized work rules (Binkley 1990; Clement 1984, 1986). Captains pilot the vessels and supervise the crew (which range from one to four on small vessels to twelve or more on large offshore vessels). Crew deploy gear (nets or lines on groundfish vessels and dredges on scallop vessels); haul, sort, and clean the catch; and store it on ice – functions that remain basic to the industry under all technologies. These skills are learned on the job from friends and relatives or other experienced fishermen, and crewing on a father's vessel remains the primary training ground for most fishermen (Thiessen and Davis 1988).

Compensation

An unusual feature of the harvesting labor market is the 'lay' system, under which vessel revenue is apportioned among capital, operating costs, and payments to labor. Under the lay or share system, labor earnings are calculated as a percentage of the value of catch, after certain vessel operating costs are deducted and shares for the captain and the capital costs of the vessel are provided (Holmsen 1972; Sutinen 1979; White 1954).[1] Earnings, therefore, reflect the productivity of the vessel and its crew (as defined by the volume and quality of the catch), the efficiency of the vessel in terms of its operating costs and crew size, and the prices of the species harvested.

Lay systems differ somewhat by vessel and port. On small, family vessels, shares may be calculated informally, while the deductions for operating expenses and the apportioning of revenue between capital and labor are formally defined on larger vessels (Kirby 1982; Doeringer, Moss, and Terkla 1986a).

In New England, the share formulas for offshore vessels are similar within ports but vary somewhat from port to port. In Atlantic Canada, most fishermen are paid in shares, although some are paid wages (DFO 1993c). While harvesters on large offshore vessels are classified as 'wage earners,' they are also paid according to share formulas and catch 'prices' that are codified in company-wide collective bargaining agreements that cover multiple ports (Kirby 1982). Surveys and our field interviews reveal that share systems in both countries can yield annual incomes ranging from a few hundred dollars for the casual harvester to over $50,000 (1990) in a good year for crew on large, offshore vessels landing high-quality fish (DFO 1993c).

Like any compensation formula that is based upon output, the lay system provides incentives for labor effort and teamwork while minimizing the need for direct supervision. The lay system is also a mechanism for determining how income and the uncertainty of catch and price will be shared within families (on kinship vessels) and between labor and capital (on corporate-owned vessels). To some extent, this dimension of the lay system insulates vessels from economic shocks and, therefore, helps to stabilize employment.

Unions

Prior to the mid-1970s fishermen in Canada were classified as 'co-adventurers,' rather than employees, and they had no legal bargaining rights (MacDonald, 1980). Although unions were voluntarily recognized from time to time by the large employers, such as National Sea Products and FPI, systematic recognition by these companies did not occur until after a major strike in Newfoundland in the mid-1970s (Clement 1984; Steinberg 1974; Kimber 1989).[2]

The offshore vessels in Gloucester, New Bedford, and Boston have been periodically unionized since the First World War (White 1954). However, although some offshore fishermen in Gloucester and New

Bedford remain represented by trade unions, the influence of collective bargaining has waned as kinship emerged as an alternative means of regulating the employment relationship (Doeringer, Moss, and Terkla 1986a). In Gloucester, where kinship practices are most firmly established, union membership in the Atlantic Fishermen's Union now has a purely social function. In New Bedford, kinship practices are coming to dominate the groundfish fleet, although half the fishermen in the port still are members of Teamsters Local No. 59.[3]

Bargaining has raised compensation levels in both countries, at least during periods when unions were strong (White 1954; MacDonald 1980), and it has affected various other management prerogatives.[4] Bargaining in Canada has also altered the structure of compensation by introducing per diem compensation guarantees, bonuses, quality premiums, and negotiations over the operation of the lay system (Binkley 1990; National Sea Products Ltd and Fishermen, Food, and Allied Workers 1991). In the case of the lay system, Canadian collective agreements specify both the share formula and a detailed structure of accounting 'prices' to be used each year for valuing catch according to species, size, and quality.

HUMAN RESOURCES PRACTICES IN PROCESSING

There are major distinctions in the employment practices of fresh and frozen processors that largely reflect the influence of technology and size of plant. Frozen production is a highly mechanized, mass production process that lends itself to larger plants. Interviews and surveys show that frozen processing plants typically employ over 100 workers (Apostle and Barrett 1987; Apostle and Jentoft 1991).

In contrast, fresh-groundfish processing is a labor-intensive, assembly-line process (involving washing, skinning and filleting, and inspection and grading for quality of fish). In Boston and New Bedford (and in Gloucester when there was a fresh processing industry), for example, fresh processing plants typically operate with a core of fifty or fewer employees (Doeringer, Moss, and Terkla 1986a; Georgianna, Dirlam, and Townsend 1993). Even in Newfoundland, where frozen processing predominates, 61 percent of processors have fewer than 100 employees and 44 percent have fewer than fifty employees (DFO 1993c). In Nova Scotia, where much more fresh fish is processed, a little over 80 percent of all fish processors employed 100 or fewer workers and 66 percent employed less than fifty workers (Apostle et al. 1985, 1992; Apostle and Barrett 1987; DFO 1993c).

Recruitment of Labor

Small fresh- and salt-fish processors often recruit their labor through kinship and friendship networks. In Atlantic Canada, for example, friends and relatives of the owner are the usual sources of labor recruitment (MacDonald and Connelly 1986a, 1986b; Ilcan 1985). Large frozen processors also recruit labor through community and kinship networks, although those that are part of multi-plant corporations may also allow transfers among plants and may resort to formal recruiting methods outside the immediate community (MacDonald and Connelly 1986a, 1986b).

Job Content and Training

Large frozen processors divide work into relatively specialized semi-skilled jobs, and work is largely routine. Jobs are less specialized in small and medium-sized salt- and fresh-fish plants, and workers are more broadly skilled because they are expected to perform different kinds of work (Apostle and Barrett 1987; MacDonald and Connelly 1986a, 1986b; Andersen and Stiles 1973; Giasson 1992). In both types of plants, skill levels are much like those in other kinds of light manufacturing and skill needs are met easily through on-the-job training.

Compensation

Hourly wages rather than piece-rates are the dominant form of compensation, although incentive pay is also used in fish-cutting operations and in the smallest processing plants (Apostle and Barrett 1987). Large processing plants pay higher wages and are more likely to operate year round than small plants because they draw upon catch from the year-round offshore fleet (in the case of Canada) or because they process fish from frozen product that can be inventoried.

These differences in wages and weeks worked translate into annual earnings differentials. In Nova Scotia, for example, annual earnings (1989) in firms with fewer than 100 employees are about 40 percent of those in firms with 250 or more employees, and the corresponding figure for Newfoundland is 32 percent (DFO 1993c).

Unions

Less than 10 percent of the processing plants in Atlantic Canada are unionized, but because they tend to be larger frozen processors, a much higher fraction of employment and production is affected by collective bargaining (MacDonald and Connelly 1986a, 1986b). Estimates for Nova Scotia suggest that over 40 percent of the processing workforce is covered by collective bargaining agreements (Apostle and Jentoft 1991). Similarly, unions and collective bargaining in New England are confined to frozen processing plants and a few large fresh processors.[5]

Interviews suggest that pay in nonunion plants in New England is roughly comparable to that in union plants of similar size, although fringe benefits are slightly better in union plants. Wage rates and fringe benefits in unionized plants in Atlantic Canada, however, are the highest in the industry, especially for women, and they set the wage pattern for the larger nonunion plants (Apostle and Barrett 1992b; Ilcan 1985; Apostle and Barrett 1987). Unions have also contributed to more formal work rules, the greater use of seniority for allocating employment, improved health and safety conditions, the introduction of grievance procedures, and enhanced job security (Apostle, Kasdan, and Hanson 1985; Lamson 1986).[6]

WORKPLACE EMPLOYMENT SYSTEMS

Employment in harvesting and processing falls into one of three systems of workplace organization – kinship capitalism, paternalistic capitalism, and corporate capitalism. Each of these employment systems is staffed according to different criteria, yields different levels of earnings, provides differing degrees of employment security, and takes different approaches to promoting labor productivity.

Kinship Capitalism

Kinship capitalism is particularly characteristic of the inshore fleet in both countries, although nearshore vessels that are family owned and some small processing facilities also operate according to kinship practices. Because boats are small, they can be crewed by immediate family or by fishing partnerships among groups of friends (Squires 1990; Doeringer, Moss, and Terkla 1986b; Jorion 1982; Barber 1992). Under kinship capitalism, employment, pay, and effort are governed by a complex set of personalized and reciprocal obligations among captain and crew, rather than by impersonal market forces (Binkley 1990; Clement 1984). Unrelated labor is hired only after kinship and friendship labor supplies are exhausted, and its employment rights are inferior to those of kinship labor (Thiessen and Davis 1988; Doeringer, Moss, and Terkla 1986b).

Kinship capitalism is also the dominant organizational form in the large-vessel offshore sector in New England (Doeringer, Moss, and Terkla 1986b). Kinship employment systems are found on Italian-owned vessels in Gloucester and on Portuguese-owned vessels in New Bedford. These systems grant family members priority access to job opportunities. Unless they possess some special skill, such as the ability to repair engines, non-Italians rarely find work on Italian vessels and non-Portuguese rarely work on Portuguese vessels. When catch levels are increasing and employment levels are high, captains of kinship vessels may complain about a 'shortage' of crewmen, but this means that they cannot find enough men whose families and friends they know – not that experienced job candidates are unavailable.

Kinship capitalism involves a commitment to provide work for relatives, even in times of substantial declines in catch and revenue. This commitment is met by maintaining the size of crews at the level necessary to keep family members employed, regardless of catch, or by rotating family members on and off boats so that available work and income are shared among all family members. Layoffs on kinship vessels are largely limited to unrelated individuals (Miller and Van Maanen 1979; Doeringer, Moss, and Terkla 1986b).

The corollary of these kinship employment preferences and obligations is the commitment of crew members to contribute to the well-being of the family production unit. High effort, smooth teamwork, and low absenteeism are expected features of the work relationship.

Corporate Capitalism

In contrast to the personal relationships of kinship capitalism, the corporate capitalism system emphasizes impersonal and often temporary employment relationships, with sharp distinctions between the authority of management and the duties of workers and an absence of ongoing employment commitments. In large-scale processing plants, corporate capitalism also involves the 'bureaucratic' employment practices of hierarchy and formalized management practices. Unions and collective bargaining are found almost exclusively in the corporate capitalism sector.

Corporate capitalism employment systems are particularly a feature of the fishing industry in Atlantic Canada, where they are found in the offshore fleet, in large processing plants, and in parts of the nearshore fleet. They are also present in those vessels of the New England offshore fleet that are staffed without regard to kinship, principally the boats in the Gloucester fleet owned by non-Italians, the New Bedford vessels that are owned by non-Portuguese, and the transient vessels from outside New England (Doeringer, Moss, and Terkla 1986b).

Although there is considerable continuity of employment in the core crews of corporate capitalism vessels, particularly on the most productive 'high-liners,' the formal employment relationship is 'employment-at-will,' with labor services being contracted on a trip-by-trip basis. The decision by individual crew members to remain with a vessel, the willingness of captains to continue to employ specific crew members, and the owners' decisions to continue their vessels in operation are governed by market considerations, rather than by family or kinship relationships.[7]

Labor productivity and efficiency on corporate capitalism vessels are the responsibility of the captain, who has widespread authority over hiring, hours of work and length of trip, and job duties. Crew size is adjusted to changes in the volume and value of catch, and vessels are taken in and out of service depending on their overall profitability. Efficiency, merit, and formal seniority are the governing considerations in staffing decisions, and job duties and assignments are more clearly defined on corporate capitalism vessels than on kinship vessels.

Corporate capitalism employment practices are also prominent in large processing facilities, especially those that are part of Atlantic Canada's vertically integrated fisheries companies. Corporate processing plants are relatively capital intensive and work is rationalized and mechanized. The pace of work is faster than it is in other types of plants, supervisory monitoring is used to ensure productivity, and workers are subject to discipline or discharge for failure to meet production targets (MacDonald and Connelly 1986a, 1986b; Apostle and Barrett 1987). Workers in this sector, however, typically receive higher pay than those performing comparable work in smaller firms (Apostle et al. 1992).

Paternalistic Capitalism

The third organizational model in the industry is paternalistic capitalism (Doeringer 1984; Apostle and Jentoft 1991; Barber 1992). Like corporate capitalism, paternalistic capitalism relies on managerial authority to achieve labor efficiency. Instead of paying high wages and using supervision and discipline to motivate the workforce, however, it adopts a low-wage, paternalistic workplace strategy that generates reciprocal effort from the workforce.

Paternalism involves establishing personalized relationships between the employer and the workforce. Management processes are not highly structured, performance standards are only loosely enforced (MacDonald and Connelly 1986a, 1986b; Apostle and Barrett 1987), and employers grant workers special favors based on need as well as performance. Workers reciprocate with loyalty and a sense of obligation to management, which results in low turnover, high effort and flexibility, and a willingness to work on short notice and for long hours during peak periods of demand (Apostle and Barrett 1992b; Giasson 1992, Barber 1992). In both countries, these paternalistic practices are found in medium-sized processing plants in smaller communities, where personal contacts can be maintained between workers and managers in both workplace and community settings.

THE TRANSFORMATION OF LABOR MARKET INSTITUTIONS

Market forces – the growth of demand for fresh fish, increasing international competition, and declines in catch – have intensified competitive pressures on the employment institutions of the industry in both countries over the last two decades. In New England, where public policy impinges lightly on markets, competition has produced only gradual change in institutions. The kinship-harvesting sector has grown in the offshore fleet, displacing both corporate capitalism vessels and the workforce protections of unionization, and there has been a trend toward somewhat greater concentration in processing. While the present stock collapse is forcing the harvesting sector to downsize, the question of the future institutional composition of a consolidated industry has not yet been addressed by either the industry or public policy.

In contrast, both competition and interventionist public policies have repeatedly reshaped the labor market institutions of the industry in Atlantic Canada. The rationalization of the corporate processing sector and the granting of quotas to the large processing enterprises, for example, were intended to make the industry competitive in global markets for frozen fish (Kirby 1982; Barrett and Davis 1984). Improving productivity and efficiency, however, led to growth in the relative importance of corporate capitalism at the expense of the paternalistic capitalism sector as low-productivity firms were closed (MacDonald and Connelly 1986a, 1986b).

A second important public policy influence on employment systems has been the goal of using the fishing industry as a source of employment of last resort, particularly in isolated rural ports where incomes are low and alternative jobs are scarce (Kirby 1982; DFO 1993c). TAC allocations and regional development subsidies are examples of public policies that have helped to stabilize the labor-intensive kinship and paternalistic capitalism sectors in Canada.

In the short term, the allocation of TACs between offshore and inshore vessels, including their affiliated processors, defines the dividing line between corporate capitalism and other forms of work organization in the industry. The TAC allocations between the two sectors, which initially gave inshore fishermen a slightly larger allocation than the offshore sector, have remained in roughly constant proportion over the past decade. By this measure, regulatory policy should have stabilized the balance among different types of work organization.[8]

What was not anticipated, however, were the incentives for investment in a nearshore fleet that would be created by these catch allocations and restrictions on entry. The phenomenon of the nearshore fleet increased the overall importance of corporate capitalism employment systems as harvesting capacity shifted away from small kinship vessels.

LABOR MARKET STRUCTURE AND LABOR EFFICIENCY

There is a considerable, and highly charged, debate over what these changes in the mix of employment systems imply for economic efficiency (Kirby 1982; Barrett and Davis 1984; Kimber 1989; Barrett 1992b). Some analysts argue that kinship and paternalistic capitalism arrangements are more efficient than those of corporate capitalism, while others assert that corporate capitalism has superior efficiency properties.

The arguments favoring smaller-scale kinship or paternalistic enterprises are based on their flexibility in organizing production, the savings from dispensing with the indirect costs of corporate bureaucracy and bureaucratic rigidities, and the efficiencies in training and capital formation that are available within family production systems. It is also sometimes asserted that intergenerational participation in the kinship sector of the industry leads to communal pressures that encourage efficient management of the fisheries resource (Stiles 1979; Byron 1976, Squires 1990; Martin 1979).

The competing view contrasts the efficiencies of scale economies and rational management practices found under corporate capitalism sys-

tems with the inefficiencies of small-scale production and the employment obligations of kinship and paternalistic employment systems. It
also points to the potential for efficient targeting of fishing effort that is
implicit in the use of enterprise quotas to regulate catch, an approach
that has been applied only to the corporate capitalism sector in Atlantic
Canada.

Comparing Efficiency Rates among Employment Systems

Economic theory presumes that this debate will be resolved by market
competition. Competition will force the replacement of less efficient
organizational arrangements by those that are more efficient, except
where there are significant market imperfections – market power, trade
barriers, subsidies, or common property rents – that can buffer inefficient organizational arrangements.

The fisheries competition between Atlantic Canada and New England
should provide a testing ground for the efficiency of alternative production and employment systems. Both countries have roughly equivalent
factor endowments, harvest similar fishery stocks using comparable
technologies, and sell undifferentiated fresh-fish commodities in the
same U.S. markets.

In face-to-face competition, the New England fresh-fish industry
would seem to have the stronger prima facie claim to efficiency. An
atomistic market structure, weak unions, the relatively open exposure to
trade, and the laissez-faire public policy of the New England industry
are characteristic of the type of highly competitive circumstances that
should weed out inefficient employment systems.

It could be concluded from the New England experience that a system
of independent harvesters and small-scale processors provides the most
efficient set of employment arrangements. Furthermore, the growth of
kinship capitalism in the New England offshore sector provides some
evidence to support the economic superiority of family-based work obligations over the market incentives of corporate capitalism in harvesting.

The contrasting experience of the Canadian industry, however,
suggests that there may be offsets to kinship advantages and the
paternalistic practices of small-scale independent processing that have
contributed to the growth of corporate capitalism. Even though the stable TAC allocations between the offshore and inshore sectors have
capped the growth of corporate capitalism in the offshore fleet, our
interviews suggest that corporate capitalism has also gained a foot-

hold in the growing nearshore fleet, which operates under the inshore quota.

Although data on the relative importance of corporate capitalism in vessels fishing under the inshore quota are not available, a rough sense of trends in the mix of employment institutions in this fleet is provided by data that distinguish between 'self-employed' independent harvesters (most of whom are presumably working on kinship vessels) and those who are 'wage earners' (harvesters outside the large processing firms who are paid wages rather than shares) under corporate capitalism. In 1981 such 'wage and salary' harvesters comprised only 8 percent of Nova Scotia's harvesting labor force and 6 percent of the harvesters in Newfoundland. By 1990, however, the comparable figures had risen to 26 percent and 12 percent (DFO 1993c, derived from table 15).[9]

Similarly, the restructuring of Canadian processing that brought about a major shift from the paternalistic capitalism of independent processors to corporate capitalism should be indicative of underlying productivity differences. The closing of obsolete processing capacity and investment in new plant and equipment under the restructuring program, for example, undoubtedly increased total-factor productivity, since production was consolidated into the most modern plants.

Accounts of this rationalization process, however, highlight the changes in efficiency derived from managerial practices, particularly those directed at increasing labor productivity. These studies report that modernized plants in the corporate capitalism sector have increased production speeds and effort levels compared with those previously prevailing under paternalistic capitalism (Macdonald and Connelly 1986a 1986b; Ilcan 1985; Barber 1992). Although higher wages, union work rules, and employee dissatisfaction with production speedups may have offset some of these efficiency improvements, there is a consensus among these studies that labor output has risen, thereby reversing the trend of declining productivity in processing in the 1970s and early 1980s that preceded restructuring (Kirby 1982).[10]

Critics of Canadian corporate capitalism attribute the growth of the corporate capitalism sector to market distortions, rather than to competitive advantage. According to this view, political influence and market power wielded by the corporate capitalist sector, not efficiency, explain why corporate capitalism survives. The rationalization of the corporate sector is seen as part of a larger pattern of public policy distortions – capital subsidies, limited entry, catch allocations, marketing ventures with foreign fishermen, and the like – that have been used to sustain a sector that has inefficient organizational practices.

This 'market distortion' thesis ignores the possible advantages of economies of scale and vertical integration, capital intensity, and the ability to target landings in the corporate capitalism sector. The analysis of subsidies by the USITC, for example, suggests that the kinship and paternalistic systems of the independent sector in Nova Scotia, rather than the large companies of the corporate capitalism sector, have been the principal beneficiaries of subsidies (USITC 1984).

In addition, there may also be imperfections in the incentives for effort and quality in the markets used by independent harvesters that can be avoided by the transfer-pricing arrangements used in the corporate capitalism sector. The production and quality incentives of the lay system for independent fishermen are set by market prices, and these prices may imperfectly estimate quality differences for catch that is sold in undifferentiated lots. In contrast, the corporate capitalism sector can fine tune its bargained lay incentives and quality premiums to approximate, or even enhance, market incentives for increasing the value of catch. This explanation would be consistent with our field interviews in Nova Scotia, showing an increased emphasis on the quality of harvests and rising premiums being paid for quality fish in the corporate capitalism sector.

Earnings and Productivity Differences

While these examples are suggestive of corporate capitalism's being more efficient than the kinship and paternalistic systems, the best indicator of the relative efficiency of different organizational arrangements would be cost accounting data from different harvesting and processing firms. Unfortunately, the few such studies that are available tend to focus on issues like overhead burdens and scale economies, while neglecting labor costs and productivity offsets (Kirby 1982; Georgianna and Hogan 1986) and they do not provide an adequate measure of efficiency differences. The one analysis of labor efficiency (Kirby 1982) concludes that the yield per pound to crew under inshore and offshore lay systems is roughly comparable, but this conclusion predates the rapid growth of fresh-fish production in Atlantic Canada and does not include other efficiencies from vertical integration with onshore production.

An alternative approach to the question of institutional efficiency is to

TABLE 5.1
Distribution of hourly wages by size of processor, Nova Scotia, 1984 (C$)

Wages	Small	Medium	Large
	(percent)	(percent)	(percent)
Less than $5	18.0	10.7	11.9
$5–5.99	30.9	38.7	22.9
$6–6.99	43.6	23.5	51.0
$7–7.99	6.3	21.5	9.7
$8 or more	1.2	5.6	4.5
Median employment	10	22	121

Source: Derived from Apostle et al. (1992).

adopt the convention of using earnings as a proxy for labor productivity. According to this metric, the Canadian evidence is contradictory as to whether kinship or paternalistic capitalism systems are more or less productive than those of corporate capitalism.

Thiessen and Davis (1988) find that crew on inshore vessels owned by captain and crew (a category that approximated the kinship sector at that time) have annual earnings that are 56 percent less than those in the corporate capitalist sector (offshore vessels not owned by the captain or crew).[11] Data from a more recent study find that 'self-employed' fishermen in Newfoundland also earn 30 percent less annually than 'wage-earning' fishermen, not including those employed by the largest off-shore processing companies (DFO 1993c, table 17-1). This same study, however, finds that in Nova Scotia self-employed fishermen (40 percent of whom earn $25,000 or more) earn 76 percent more than wage-earning harvesters (DFO 1993c, table 17-2).

None of these annual earnings differentials, however, takes account of differences in days at sea between self-employed and wage-earning fishermen. The Nova Scotia data also do not reflect possible differences in the mix of species caught and/or in the distribution of catch to fresh and frozen markets between self-employed fishermen and crew who are paid wages.

Earnings data for Canadian processing (table 5.1) are more consistent, showing that large corporate capitalism plants (defined as large processors with a median employment of 121) have higher wage scales than either medium-sized (median employment twenty-two) or small (median employment ten) processors (Apostle et al. 1992). Almost half the employees in small and medium-sized plants (the size categories that correspond most closely to the paternalistic capitalism sector) earn less than C$6 per hour (1984), compared with only about one-third of the employees in large plants. These differences persist even after the sex composition of the processing labor force is controlled for.

INSTITUTIONAL ADJUSTMENTS TO ECONOMIC CHANGE

The different employment systems operative in the fishing industry also affect its dynamic performance in the face of changing economic conditions. Each system has its own patterns of response to change and its own margin of adjustment. These different adjustment patterns have implications for both income distribution and efficiency.

Earnings and Employment

The lay system of pay in harvesting immediately translates changes in catch and price into changes in earnings for crew. Because the kinship system retains its labor, however, changes in landed value are shared among a workforce that adjusts quite slowly to changing levels of catch. Only workers without kinship connections are quickly affected by variations in catch. In both countries, this sector is insulated from the employment effects of the catch declines in the industry, but labor earnings will fall unless there are offsetting price increases.

Conversely, the corporate capitalism harvesting sector adjusts its labor force relatively quickly to changes in catch levels while sharing changes in catch value among those who are employed. Periods of declining catch, therefore, translate into falling employment for the corporate capitalism workforce in both countries.

Under the lay system, however, if declining catch leads to higher fish prices, there could be higher incomes for those fishermen who remain employed. Aggregate data on earnings in Canada indicate rapidly rising real incomes among harvesters as catch value rose in the mid-1980s followed by sharp declines after 1987. National Sea Products, for example, asserts that annual earnings of its offshore crew averaged $30,000 in good years compared with $20,000 in bad years (Kimber 1989).

This pattern is also confirmed by our interviews in Nova Scotia and New England. In Gloucester and New Bedford during the early to mid-1980s earnings of offshore fishermen exceeded those of manufacturing workers by a factor of two to three (Doeringer, Moss, and Terkla 1986a) before declining as catch fell and prices stabilized. Although earnings from groundfishing have declined relative to other occupations, recent interviews suggest that offshore scallopers in New Bedford can still earn $40,000 to $50,000 a year.

With the exception of the small, kinship, processing firms in Atlantic Canada, the processing sectors in both countries adjust labor demand to changes in the volume of catch. Since wage rates are generally tied to local labor markets rather than to fish prices, wages are not directly affected by declines in catch. Prior to the recent stock collapse, however, the rationalization of processing and the consolidation of plants had both reduced employment and increased annual earnings in Atlantic Canada's corporate capitalism sector because of higher wages and some increase in the proportion of year-round jobs in the industry (Apostle and Barrett 1992b, 1992c).

'Sticky' Labor and Allocative Efficiency

A second dimension of the institutional adjustment process involves the extent to which labor inputs to the industry are sufficiently flexible to allow for the efficient allocation of labor between the fishing industry and other sectors of the economy. Where labor inputs are completely variable, economic change will result in the rapid reallocation of labor to its most productive uses. Where labor inputs are relatively inflexible and mobility is imperfect, surplus labor, which could be employed more productively elsewhere in the economy, is retained in the industry.

Even though harvesting employment varies somewhat with catch and processing tends to rely on flexible and 'on-call' labor, fishing industry labor markets have long been characterized as relatively inflexible. Surveys of fishermen routinely report a strong commitment to fishing as 'a way of life,' and inshore fishermen tend to live in small, rural ports, where there are few employment alternatives and where attachment to occupation and community are thought to be highest (Apostle, Kasdan, and Hanson 1985; Pollnac and Poggie 1988). Workforce attachment is also reinforced by the worksharing arrangements of kinship capitalism, as has been demonstrated for kinship vessels in Gloucester and New Bedford (Doeringer, Moss, and Terkla 1986b) and by work rotation under paternalistic capitalism.

Unemployment insurance payments contribute further to this attachment by subsidizing seasonal incomes from fishing and processing. In New England, inshore fishermen participate in the fishery around twenty-five to thirty weeks per year, while offshore fishermen have traditionally worked forty to forty-five weeks per year (Doeringer, Moss and Terkla 1986a). Seasonality is even greater in Atlantic Canada, where studies indicate that the average full-time fisherman experiences about twenty weeks and the average part-time fisherman about twenty-six weeks of unemployment per year.

In New England, unemployment insurance eligibility is mainly limited to fishermen employed on offshore vessels and on the larger inshore vessels who fish more or less year round (fishermen on small inshore vessels are excluded from eligibility because they are considered independent contractors). Lower payment levels and coverage, when compared with those of Atlantic Canada, are apparent in the unemployment insurance data. For example, one study estimated that Massachusetts fisherman received unemployment insurance payments in 1981 equivalent to about 6 percent of the total wages and salaries in covered employment in fishing (Doeringer, Moss, and Terkla 1986a).

The regular Canadian unemployment insurance program applies to 'wage earners,' largely fishermen employed in the offshore fleet and processing workers (Kirby 1982; Employment and Immigration Canada 1991; DFO 1993c). There is also a special unemployment insurance program for self-employed inshore and nearshore fishermen, however, which covers the bulk of the fisheries workforce (Employment and Immigration Canada 1991). The latter program is particularly generous when it is compared with its U.S. counterpart (Kirby 1982). Eligibility for this program can be established after only ten to fourteen weeks of work (as opposed to twenty weeks under the regular unemployment insurance program), six of which must be in fishing (Employment and Immigration Canada 1991), and a recipient can receive up to twenty-seven weeks of benefits equal to 57 percent of insured earnings (DFO 1993c).

One study found that over 60 percent of the fishermen covered by the special fishermen's unemployment insurance program received benefits (Gardner Pinfold Economic Consultants Ltd 1986). The Kirby Task Force reported that unemployment insurance constituted 16 percent of the pretax income of full-time fishermen and about 13 percent for part-time fishermen (Kirby 1982).

More recent data show that unemployment insurance payments to fishermen during the period 1985-9 averaged 26 percent of fishing income in Nova Scotia and 92 percent of fishing income in Newfoundland (table 5.2), and there has been a steady upward trend in the fraction of income from unemployment insurance for the fishing industry labor

TABLE 5.2
Sources of fishermen's income in Nova Scotia and Newfoundland: 1981, 1985–90 (C$)

	Nova Scotia				Newfoundland			
	Total income from fishing	Unemployment benefits as percentage of fishing income	Total income from all sources	Fishing income as percentage of total income	Total income from fishing	Unemployment benefits as percentage of fishing income	Total income from all sources	Fishing income as percentage of total income
1981	$11,100	18	$16,900	66	$2,800	96	$8,800	32
1985	16,125	26	22,392	72	4,393	106	10,478	42
1986	21,227	22	28,081	76	6,181	86	12,586	49
1987	24,138	20	31,496	77	8,946	63	15,797	57
1988	18,288	30	26,684	68	7,368	89	15,377	48
1989	16,734	34	25,101	67	6,021	114	14,325	42
1990	14,900	38	26,200	57	4,300	160	14,700	29

Sources: Revenue Canada, unpublished data, for 1985–9; DFO 1993c, for 1981, 1990.

force over the decade of the 1980s. By 1990 the ratio of UI benefits to employment income for self-employed fishermen in Nova Scotia was 31 percent and unemployment benefits in Newfoundland actually exceeded employment income by 8 percent (DFO 1993c, tables 24-1, 24-2). The magnitude of these benefit levels is likely to encourage the attachment of labor to the industry (Ferris and Plourde 1982).

Processing employment is similarly affected by fluctuations in catch. In Atlantic Canada, for example, a typical plant operates for six months or less each year (Barrett and Davis 1984, MacDonald and Connelly 1986a, 1986b; Apostle and Barrett 1992b) because of seasonality in landings. As landings have fallen, the ratio of UI benefits to earned income for processing workers has also risen – from 28 percent in 1981 to 56 percent in 1990 in Newfoundland and from 13 percent to 22 percent in Nova Scotia (DFO 1993c, tables 24-1, 24-2). Paternalistic practices of work rotation are often deliberately linked to unemployment insurance as a way of securing labor reserves for the processing industry, but such practices may be weakening in Atlantic Canada as corporate capitalism replaces paternalistic capitalism.

SUMMARY

The employment relationships in fishing and processing are regulated by a number of distinctive institutional arrangements governing labor recruitment, compensation, assignment, job security, and the organization of work. Some of these institutional differences stem from unions and collective bargaining in large-scale enterprises, but many are the result of employer strategies or of informal customs and norms developed within work groups in small and medium-sized enterprises.

Each of these arrangements – kinship capitalism, paternalistic capitalism, and corporate capitalism – also has different consequences for labor productivity. Available data on productivity and earnings, as well as field interviews, suggest a pattern of systematic efficiency differences by type of employment system and product. The highest value-added in harvesting is found among kinship vessels that harvest primarily high-quality fish for fresh markets, followed by corporate capitalism vessels that serve frozen and fresh markets; kinship vessels serving frozen markets have the lowest value-added per harvester. In processing, the value-added is highest in the corporate capitalism sector for both frozen and fresh products.

In addition, each employment system imposes its own specific constraints on labor adjustment, which affect both the welfare of the work-force and the efficiency of the industry. Kinship systems stabilize employment and share income, paternalistic systems rotate jobs and unemployment, and corporate capitalism has rigid wages and flexible employment. In the aggregate, these constraints result in multiple margins along which the industry responds to change.

6

Economic Institutions and Fisheries Trade

In the preceding chapters we saw how the fishing industries of New England and Atlantic Canada evolved into two distinct production systems. The efficiency of each system is linked to the distinctive mix of institutions – organizational arrangements, regulatory practices, industrial policies, and market niches – existing in each country.

The New England system is essentially one of many independent fishermen and small processors concentrating largely on specialized markets for fresh fish and operating under what has historically been a relatively laissez-faire regulatory regime. The Canadian system has a similar atomistic fresh-fish sector, but it also has a highly concentrated and highly regulated large-scale processing and harvesting sector. This concentrated sector has traditionally specialized in frozen-commodity production but now has expanded into U.S. fresh-fish markets.

There is widespread agreement that efficient regulation is a precondition for efficient production under both systems. It is also apparent that neither country has been able to manage its fisheries stocks in a way that achieves maximum economic yield. In addition, most analysts would agree that mass production in processing is efficient for frozen-commodity markets, although the extent to which market concentration is inevitable is less clear.

There is considerable controversy, however, over the competitiveness of alternative production systems in fresh-fish markets, particularly for mass retailing of fresh fish, the market niche that has the greatest potential for substantial growth. Does the flexibility of the atomistic system of independent fishermen and processors result in more efficient production than the bureaucratic system of large-scale enterprises, or does vertical integration and the economies of scale (particularly when coupled with enterprise quotas) of the large-scale sector more than offset bureaucratic rigidity?

One approach to resolving this question is the comparison of production costs under the different systems, but such studies are relatively rare and their results are inconclusive. For example, research on the Canadian processing industry conducted for the Kirby Commission indicates that large, corporate processors may be more bureaucratic than smaller firms, but that the overhead burden per unit of sales is somewhat lower in the larger firms (Kirby 1982). This finding is consistent with scale economies in management and marketing. On the other hand, studies of New England fresh-fish processing suggest that there are relatively constant long-run marginal costs and few scale economies at the plant level, and the Kirby report found that there were not large differences in unit costs between alternative organizational arrangements in the Canadian processing industry (Georgianna and Hogan 1986; Kirby 1982).

To complicate the picture further, all of these studies emphasize the importance of factors such as financial capital and excess capacity as determinants of business performance. For example, the most ambitious assessment to date of the Canadian industry's performance concludes that there were too many independent fishermen and too much obsolete processing capacity in the early 1980s as a result of policy pressures to preserve jobs (Kirby 1982). Further exacerbating these problems during the early 1980s was a level of short-term debt financing in the corporate capitalism sector that could not be sustained during periods of high interest rates, depressed markets for frozen fish, and falling catch levels. These factors tend to obscure the effect that differences in scale, ownership, and employment systems would have on an industry free of government industrial policy initiatives and operating under more stable financial circumstances.

Direct performance comparisons between production systems are limited by the lack of comparable data. The processors studied for the Kirby report were generally large and were more often committed to producing frozen, as opposed to fresh, fish, whereas the New England studies involved only small and medium-sized fresh-fish processors. Similarly, comparisons of groundfishing costs among different size classes of vessels may not capture the effects of differential flexibility in the ability to switch to other species such as lobster. In addition, since New England and Canadian processors draw upon raw product harvested under different regulatory regimes, cost comparisons are further inhibited.

Given the limitations of these cost analyses, an alternative approach to testing the efficiency of the different production arrangements is to examine how each system fares in the arena of international competition. In this chapter recent trends in fresh-groundfish trade between the United States and Canada are examined and an institutional model to explain these trends is developed. This model is then applied to fresh-fish trade data to demonstrate how differences in industrial structure, industrial policy, regulatory policy, and economic development concerns between the United States and Canada translate into a 'bottom line' effect on international competitiveness and trade.

PATTERNS OF FISHERIES TRADE

During the 1970s and early 1980s around 80 percent of all groundfish consumed in the United States was imported, primarily from Canada (USDOC 1985; Statistics Canada, various years). Almost all of the Canadian groundfish exported to the United States originated in Atlantic Canada and were sold through processing and distribution channels in the northeastern United States. From the Canadian viewpoint, the United States represents a profitable market that consumes almost 60 percent (by weight) of Canadian groundfish products (Kirby 1982; USITC 1984). This fisheries market is highly segmented by product lines – fresh, frozen, and salted – and these segments have important effects on U.S. and Canadian trade flows.

Canada has long been a major supplier of frozen-groundfish markets in the United States, presumably reflecting the cost advantages it enjoys in this market sector. Exports of fresh whole and filleted Canadian groundfish to the United States, however, were limited until the late 1970s, when they began to increase rapidly. These exports peaked in the 1986–7 period at around eight times their level in 1977 (figure 6-1; table A-1), and even after the recent sharp decline in Canadian landings, they remained almost two and a half times their 1977 level (Statistics Canada 1977–90; U.S. Bureau of the Census 1991–2, 1993).[1] Fresh-groundfish exports have increased from less than 9 percent of total Canadian groundfish exports (by weight) in 1977 to over 40 percent in 1993 (table A.15).

Frozen- and Salt-Fish Trade

Although the trade in frozen- and salt-fish products has always been in one direction (from Canada to the United States) and has not changed

FIGURE 6.1

Canadian fresh-groundfish exports to the United States: 1977–93

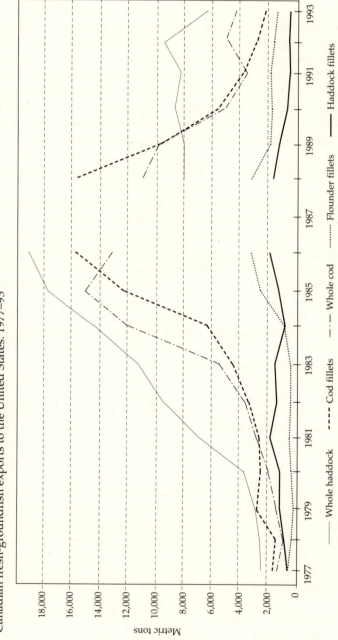

Metric tons

Sources: Statistics Canada, various years, 1975–86; Statistics Canada, International Trade Division, unpublished data, 1988–90; U.S. Bureau of the Census 1991–2; U.S. Bureau of the Census 1993. Data for 1987 are not available.

— Whole haddock — – – Cod fillets — · — Whole cod .... Flounder fillets —— Haddock fillets dramatically over the last decade, this segment of the market has an important bearing on Canadian fresh-fish exports. On an annual basis, the aggregate supply of fish to Canadian processors is almost totally inelastic because of strict management quotas that do not respond readily to changing prices, exchange rates, or costs. Since quotas control the total quantity of the harvest available to processors, their processing decisions focus on whether to freeze, salt, or sell fresh. While differences in quality limit the extent to which catch can be allocated to different end markets, the quantity of fresh-fish exports is somewhat dependent on prices in the frozen-fish and salt-fish markets, as well as on those in fresh markets.

Almost 60 percent of salt-fish production takes place in Nova Scotia (DFO, *Annual Statistical Review*, 1987; Statistics Canada 1986). Because Nova Scotia is also the major source of Canadian fresh-groundfish exports, salt-fish prices could play a role in processors' decisions to export fresh fish, particularly small salt-fish processors in southwest Nova Scotia who do not have the option to freeze fish. The lower value of salted products and the fact that only around 20 percent of Canadian salt-fish production is exported to the United States, however, suggest that salt-fish processing is not a significant factor in fresh-fish exports.

Frozen-fish exports represent a much more significant alternative to fresh-fish exports. Most of the raw material input of U.S. frozen-fish products is imported, and the United States represents the major market for Canadian processed frozen-fish products.[2] The bulk of Canadian frozen blocks and fillets are purchased by several large U.S. food conglomerates for processing into fish sticks or packaging for sale in supermarkets. These firms are much larger than the Canadian processing firms, and buyer concentration at the wholesale end of the U.S. frozen fish market has limited the opportunities of Canadian frozen-fish companies for direct marketing to retail establishments (Kirby 1982).

The U.S. frozen-fish market is divided into different segments according to the quality of fish required by buyers. The lower the quality of fish required, the more price sensitive the buyer tends to be and the less profitable the market for the supplier (Kirby 1982). The lowest-value segment uses primarily frozen blocks, which are then processed further into fish sticks and the like. Institutional food services (such as schools, hospitals, and cafeterias) and some retail supermarkets represent the middle-quality segment of frozen-fillet and block products. The highest-quality frozen fish are purchased (as fillets) by restaurants and quality-conscious retail supermarkets.

Traditionally, frozen blocks have been a low-profit product line for the Canadian industry. The average real unit value (total value of exports divided by total quantity, henceforth referred to as 'average price') of frozen-block exports moved cyclically with the U.S. economy over the 1977–90 period, falling by 5 percent overall, and a similar declining trend is found for specific species, such as frozen-cod blocks (table A.16). Despite this decline in real average price, which, ceteris paribus, should have resulted in a shift toward a more profitable export product, the quantity of Canadian frozen-block exports to the United States increased by 43 percent over this period (table A.15). Exports declined dramatically in the early 1990s, however, coincident with the rapid reduction of Canadian groundfish stocks, particularly the closure of much of the Newfoundland cod fishery, and a 16 percent decline in the average price of frozen blocks in 1993 (tables A.15, A.16).

Almost all of the Canadian frozen-fillet exports to the United States enter the medium-quality frozen-fillet market. Canada is a dominant supplier of frozen portions to the institutional market (schools and the military) but is only a secondary supplier to national food chains – the higher-value end of the frozen market. Frozen-fillet exports increased by a little over one-third during the 1977–90 period (table A.15), although there was a 5 percent decline in the real value of frozen fillets. Like the pattern of frozen-block exports, frozen-fillet exports have declined dramatically since 1990 (table A.15), although the average price of frozen cod fillets rose by almost 18 percent between 1990 and 1993 (table A.16).

Fresh-Fish Trade

Canada has been, and continues to be, the only significant foreign supplier of fresh whole and filleted cod to the United States.[3] In contrast to the more gradual shifts in the level, composition, and value of the frozen-fish trade until 1990, fresh-fish exports to the United States grew rapidly during the 1980s (figure 6.1; table A.15).

In 1979 Canadian exports of fresh whole groundfish equaled only 4 percent of U.S. production (landings), and exports of fresh fillets represented only 13 percent of U.S. fillet production. By 1985 these exports had risen to 28 percent and 20 percent, respectively, and Canadian cod fillets accounted for almost half the volume of U.S. production in 1984 (USITC 1984, 1986; Schrank, Tsoa, and Roy 1987). While fresh-fish exports began to decline in the late 1980s and early 1990s as Canadian landings fell, they declined at less than half the rate of frozen exports, and the overall increase in exports over the last fifteen years was substantial (figure 6.1; table A.15).

What Has Caused the Trade Surge?

Traditional economic models of international trade explain this changing pattern of fresh-fish trade by factors related to prices and factor costs. Changes in landed values, exchange rates, tariff protection, or transportation and other input costs that reduce the import prices of Canadian fish relative to those in the United States should expand fisheries trade flows from Canada to the United States and vice versa. Changes in relative factor costs do not appear to have been substantial enough, however, to account fully for either the dramatic increase in Canadian exports during the late 1970s and early 1980s or the falloff in the late 1980s. In addition, the timing of these increases and decreases is also inconsistent with traditional cost-based explanations.

This divergence between the trend in underlying economic variables and changing trade flows has prompted a number of alternative explanations of the trade flow pattern, several of which have been advanced. For example, one explanation is that the increase in U.S. demand, coupled with declining U.S. landings and the appreciation of the U.S. dollar, altered relative prices in favor of Canadian exports (USITC 1984). Another is that U.S. processors import Canadian fish to counterbalance exogenous shortfalls in U.S. catch in order to maintain full-capacity utilization (Hogan and Georgianna 1989). A third is that Canadian government subsidies are triggering trade flows (USITC 1984). As shown in the following sections, however, these explanations, like the traditional trade model, do not appear to be wholly consistent with the evidence.

Market Prices

Since Canadian processors face an inelastic overall supply of fish because of catch quotas but can choose how to allocate this supply among different products to export, price-based theories would predict that significant increases in prices would lead to increases in exports of particular product lines. Contrary to this prediction, the real average price of all Canadian groundfish exports was falling over the entire 1978–84 period, when most of the expansion in fresh-fish exports took place.

Cod is the dominant species in all export markets. The cod products with the largest declines in real average prices during the trade surge showed the largest increases in exports. During the 1978–84 period, the real average price of whole cod exports declined by over 26 percent (table A.16). In contrast, the real average price of fresh, cod-fillet exports fell less rapidly, declining by 24 percent during the same period. Yet whole-cod exports expanded at a much more rapid pace than cod fillet exports (figure 6.1; table A.15). Likewise, while the real average price of frozen cod fillets (the major alternative market outlet for cod) was declining much less rapidly (falling by 14 percent during this period), exports of this product to the United States increased only 64 percent, in contrast to the more than thirteenfold increase in whole-fish exports during this same period. Although real average prices rose or stayed constant during the 1985–90 period, exports of fresh and frozen cod declined.

Declines in U.S. Landings

There is no consistent pattern of Canadian fresh-fish exports replacing U.S. landings. During the 1982–4 period, when the largest increases in whole-cod exports were occurring, U.S. landings declined by 7 percent, and more recently (1985–90), when U.S. landings increased by 16 percent, exports of whole cod declined, as this 'substitution' hypothesis would suggest. Canadian whole-cod exports began to increase rapidly in the 1978–80 period, however, while U.S. landings were also increasing substantially (36 percent) (figure 6.1; tables A.2, A.15). Thus, while it appears that U.S. and Canadian landings influence Canadian fresh-fish exports during some periods, other factors must also have triggered the surge in Canadian fresh-fish exports during the late 1970s and early 1980s.

Exchange Rates

Exchange rates often have a major effect on trade, but their relationship to fresh-fish trade has been erratic. The Canadian dollar actually appreciated by 1.3 percent in real terms when exports of fresh-fish were increasing most rapidly during the late 1970s and early 1980s (USITC 1984). While the 5 percent real depreciation between mid-1983 and early 1985 could have amplified the trend in fresh-fish exports, it does not appear to have been substantial enough to have been the catalyst for the initial surge of fresh-fish exports.[4] Between 1986 and 1990, however, the appreciation of the Canadian dollar (16 percent) was large enough to have contributed to the decline in Canadian fresh-fish exports during this period.

Subsidies

Production subsidies can distort trade patterns, and there is ample evidence of extensive Canadian government subsidies to the fishing industry (USITC 1984; Corey and Dirlam 1982; Charles River Associates 1982; USDOC 1986c). One study concluded that subsidies for vessel construction and loans alone were equivalent to almost 20 percent of the price received by Canadian importers (Corey and Dirlam 1982). Most of the subsidy programs have been in effect since the 1950s and early 1960s, however, and there is no evidence that the use or level of these subsidies increased substantially during the 1980s when Canadian fresh-fish exports were increasing. On the contrary, some were being phased out because of recognized excess capacity in the harvesting and processing sectors.

Subsidies also should have affected the flow of Canadian whole-fish and fillet exports. Yet the USITC investigation found only a small subsidy effect on whole fresh fish and concluded that there was no effect of subsidies on filleted fresh fish (USITC 1984). Thus, there is little evidence to suggest that subsidies could have induced the substantial changes in fresh-fish exports.[5]

Import Duties

Similarly, there is no evidence that import duties have been a major factor affecting the quantities of Canadian fish exports. Import duties have been very low and have remained substantially unchanged for many years. Moreover, some fresh-fish products – whole cod and flounder fillets – were duty free until January 1986, when a countervailing duty of 5.82 percent was placed on fresh, whole Atlantic Canadian groundfish.[6] While this tariff may have contributed to the decline in whole-cod exports thereafter, field research with processors revealed that any impact was marginal at most.[7]

Transportation Costs

Canadians remain at a comparative disadvantage in the New England regional market, since their fish are, at a minimum, one day older than locally caught fish when they arrive by truck in Boston. Interviews with

National Sea Products, however, indicate that fresh-fish sales had been expanding gradually in the 1980s from the small, regional, New England market to the larger markets of the midwestern sections of the United States. Until declining stocks forced a cutback in fresh-fish production, much of this expansion was generated by improved technology for extending the shelf life of fresh-fish fillets and by improved air transportation networks to key inland U.S. markets.

These developments narrowed the locational disadvantage of Canadian producers relative to New England processors, since air transport costs to these regions are not much higher from Nova Scotia than from Boston and travel time is about the same. The penetration of inland markets was also aided by New England catch levels that were barely sufficient to satisfy the regional market, thereby leaving the expanding markets to Canadian supplies. Overall tranportation costs have not changed sufficiently, however, to explain the surge in Canadian exports to the United States.

INSTITUTIONAL MODELS OF FISHERIES TRADE

Although each of these factors may partly explain patterns of fresh- and frozen-fish exports from Canada to the United States during certain time periods, they do not account fully for the substantial surge in Canadian fresh-fish exports between 1977 and 1985. Our interviews with industry representatives indicate that the key to a full understanding of this surge lies outside the traditional explanations of trade. The changing roles of economic institutions and public policies during this period, as well as conventional economic forces, must also be taken into account.

This view is consistent with recent advances in trade theory that have highlighted the importance of industry structure – monopolistic competition, differentiated products, and increasing returns – as possible determinants of trade flows (Krugman 1989). Realization of this importance has led to theoretical trade models in which strategic business decisions and competitive, rather than comparative, advantage govern trade flows in some industries. While this literature rarely focuses on specific institutional arrangements, it is clear that once the assumptions of constant returns to scale and perfect competition are relaxed, factor costs no longer solely determine unit costs and unit prices.

Although groundfish products are relatively standardized, other elements of the industry – high concentration, increasing returns to marketing and distribution in the corporate sector in Atlantic Canada, and considerable 'learning by doing' in marketing – are indicative of monopolistic competition. Its presence points to the potential importance of formulating institutional models to explain international patterns in fisheries trade.

The Monopsony Model

One such institutional model is briefly sketched in the USITC (1984) analysis of the Canadian fresh-fish industry. This model posits that the increase in the level of concentration in Canadian processing following the restructuring of the industry in 1984 reinforced the monopsony power of large processing firms in buying fish from the independent fishermen and processors who are the source of much of the fresh whole fish that is exported to the United States.

Such market power can affect the level of Canadian fresh-fish exports in one of two ways. One possibility is that large firms buy fresh fish directly from independent fishermen at monopsonistic prices. In contrast, the low concentration of processors in the U.S. market makes it unlikely that they have substantial monopsony power, and therefore they must purchase fish at market prices. The difference between monopsonistic and market-based ex-vessel prices would give large Canadian firms a cost advantage in exporting fresh fish to the United States that would translate into increased exports.

Alternatively, large firms could exercise their monopsony position to depress the price of fish destined for the frozen and salt-fish markets. When faced with the choice of selling fish to large processors for freezing or salting or exporting fresh fish to the United States at market prices, independent Canadian processors and distributors choose to ship more fresh fish to the U.S. market than they would if all prices were set competitively.

While the presence of a few dominant buyers in Atlantic Canada raises the possibility of monopsony power, there is little independent evidence to support this model as an explanation of changing trade patterns. The increases in fresh whole-fish exports were already substantial prior to the restructuring of the large processing firms. In addition, although exchange-rate-adjusted ex-vessel fish prices are lower in Canada relative to those in the United States, this fact appears to reflect the lower opportunity cost of Canadian fisheries labor and the high seasonality of inshore landings in Canada – considerations that are independent of market structure (USITC 1984).

Moreover, even if ex-vessel price differentials are caused by monopsony power, lower Canadian ex-vessel prices are not particularly evident in U.S. fresh-fish markets. Discussions with processors suggest that daily differences in the prices New England dealers pay for New England and Canadian whole fish are not large and represent quality differences in freshness and size, rather than differences in ex-vessel prices (Mazany, Barrett, and Apostle 1987).

The Industrial Structure Model

A second, more realistic, institutional model focuses on industry structure. Extensive interviews conducted with processors and distributors in both countries indicate that the changing industry structure, the evolution of marketing channels, changes in management regulations, the development of new markets, and changing technology have increased the profitability of Canadian fresh-fish exports to the United States. These factors, along with changes in traditional cost factors, are the principal forces behind the changing patterns of U.S.-Canadian fresh-fish trade.

This industrial structure model divides the Canadian fresh-fish export market into two sectors – one representing the market for fresh fillets and the other the market for whole fish. The fresh-fish fillet sector is composed primarily of National Sea Products and a few other Canadian processors in southwest Nova Scotia that are large enough to achieve economies of scale in all product lines. These firms sell most of their fish in the world frozen-fish market as fillets or blocks, and they are also active in the world salt-fish market, both markets that are highly competitive. Although they may sell some fresh whole fish, most of their fresh product is filleted.

Since the bulk of their product is sold in frozen markets, prices in these markets serve as reference points against which to compare fresh-fish market prices. With the option to freeze always present, large processors are able to set prices for their fresh fillets at a level that will make a switch from frozen to fresh profitable for them.

There have been several institutional changes over the last fifteen years that have had an even stronger impact on fresh-fillet exports from this sector than that of traditional price considerations. The restructuring of the processing companies in the early 1980s increased the size of National Sea Products, expanding its control over a larger portion of existing fish supplies. At the same time, the introduction of enterprise quotas allowed large Nova Scotia processors to plan their harvesting

operations in order to control the stability and predictability of this supply much more effectively than either the small Canadian processors or
the atomistic New England processors (all of whom must rely on uncertain daily landings and purchases of fish from brokers at uncertain
prices in order to meet large orders).

Altogether, these changing institutional factors help to explain, in
part, the rapid increase in fresh-fillet exports to the United States. As
these changes are all fairly recent, however, and because U.S. mass markets for fresh-fish outside New England are still evolving, the full force
of their impact on the export market has not yet been felt.

Growth in mass markets for fresh-fish fillets among large supermarket
and restaurant chains requires a continuous supply of high-quality fish.
In terms of supply, Canadian processors have always had an advantage
over their American counterparts because of their access to a much larger
resource. This advantage was reinforced by the restructuring of the large-
enterprise sector and the establishment of enterprise quotas so that this
sector could begin to penetrate the growing market for fresh fillets.
Further expansion into this market could enable the large Canadian processors to bypass the wholesalers and brokers (concentrated in New
England) that currently distribute most of the Canadian product, thereby
further enhancing the competitiveness of these large firms.

The Canadian government's efforts to aid large processors through a
combination of enterprise quota allocations and mergers to achieve
economies of scale needed for successful competition in world frozen-
fish markets has been both an asset and a liability in entering the fresh-
fish market. Scale economies and predictable levels of output are major
advantages in the mass marketing of fresh fish. Because the Canadian
industry has been oriented toward the less quality-conscious frozen
block and fillet markets, however, it did not historically develop the harvesting expertise or production processes needed to ensure the consistently high-quality fresh product demanded by mass markets in the
United States.

Consequently, the large processors' capacity to respond to changing
frozen-fish prices and to take advantage of their growing competitive
advantage in the fresh-fillet market has been limited by the amount of
high-quality fish they have been able to produce. During the 1970s and
1980s quality problems tended to offset many of the advantages of scale
economies and predictability in production in the large-enterprise sector
(Kirby 1982). Canadians are aware of this problem and have been introducing improvements in the handling and shipping of fillets in an
attempt to correct it. Time to market and limited experience in selling

directly into high-quality markets, however, are factors that continue to put large Canadian firms at somewhat of a disadvantage when competing against U.S. processors.

The whole fresh-fish export sector is composed primarily of the many small and medium-size processors and distributors in southwest Nova Scotia. Rising demand and increased prices for fresh fish, relative to those in alternative markets for salted and frozen fish, have encouraged the growth of this sector. Institutional forces connected with marketing networks have been important in this market as well.

In contrast to the large firms, most of the smaller processors cannot negotiate direct-sales contracts with large U.S. fresh-fish buyers because they cannot guarantee a predictable supply of high-quality fish. Therefore, they must rely on New England brokers and wholesalers to market their fresh-fish. Often they will send their fish on consignment via local brokers, who truck the fish to Boston and sell them to U.S. brokers.

These exporting relationships developed by suppliers in southwest Nova Scotia have expanded over the last decade, as have counterpart initiatives taken by New England buyers to establish regular buying arrangements with specific Nova Scotia processors. Neither the relationships built by Canadian suppliers nor those developed on the demand side of the U.S. whole-fish market, however, can match the level of predictability of fresh-fish supplies to their customers offered by the large Canadian processors.

The composition of increasing trade flows from Canada to the United States is therefore made up of two distinct economic segments: (1) fresh (mostly whole) fish supplied primarily by independent fishermen and processors in Nova Scotia to the existing New England network of processors and distributors, and (2) fresh fillets supplied mostly by large-scale processors in Nova Scotia directly to mass market outlets in the United States or through New England distribution channels.

The large-scale processors are developing a competitive advantage through their marketing scale economies and their ability to guarantee a large supply of product at a specific time and at a specified price. In contrast, the small, whole-fish exporters are essentially price takers, choosing whether to subcontract with the larger Canadian processors or to sell their fish directly into the New England wholesale market.

The large Canadian processors have an interest in reducing the flow of fresh-fish exports from independent processors because the competition depresses the prices they can negotiate with large American buyers. If a large quantity of southwest Nova Scotian whole fish is sent to New England, prices will be lower and orders from large buyers will fall.

Therefore, the large Canadian companies should be motivated to contract with the independents to market their fish – freezing the fish when the fresh-fish prices are low and negotiating direct contracts with retailers or shipping fillets to U.S. distributors when prices are high. Consequently, the economic incentives facing the large Canadian processors seem to be moving them away from using whatever monopsony power they might have possessed toward offering competitive fish-brokering services to the independents.

EVALUATING TRADITIONAL AND INSTITUTIONAL EXPLANATIONS

The industrial structure model described above predicts that there will be growth in the level of fisheries trade in favor of Canada, independent of the underlying fundamentals of landings and fish prices in the two countries. It also predicts that the composition of these exports should shift from independent suppliers to the large-scale processing sector.

Because this model presupposes the existence of relatively healthy fish stocks, it cannot predict trade flows during unusual periods of stock collapse. The model can be applied, however, to periods of relatively abundant stocks, such as those characterizing both countries in the late 1970s and 1980s, to evaluate the influence of institutional factors on Canadian exports over and above that of traditional economic factors. Since the forces driving the model are not likely to change, the model should also apply once stocks return to healthy levels.

While the full complexities of this model cannot be tested, its consequences for overall trade flows can be evaluated through a relatively simple econometric model. The basic model uses the conventional estimation of fresh-fish export supply and demand as functions of fish prices, landings, and exchange rates (Crutchfield 1985; Felixson, Allen, and Storey 1986; Tsoa, Schrank, and Roy 1982; Hogan and Georgianna 1989). Industrial structure is built into this model through a separate estimation of supply and demand for fresh fillets and whole-fish exports and an introduction of a time trend to capture the institutional influences of organizational learning identified in our field research.[8]

The Model for Whole Fish

In the whole-fish export model, it is assumed that the demand for Canadian fish by U.S. processors and wholesalers is influenced by traditional economic factors – the U.S. wholesale price of fresh fish, the U.S. ex-vessel price of fish, the price of Canadian whole-fish and fillet exports, and the volume of U.S. landings, as well as by the seasonal nature of fish consumption in the United States (there is a large increase in demand during the spring associated with the celebration of Lent). This traditional trade model is then modified by the introduction of a time trend as a proxy for the influence of shifts in consumer preferences and the development by New England processors and distributors of buying relationships with suppliers of whole fish in southwest Nova Scotia.

On the supply side, the model also assumes Canadian exporters of whole fish are motivated by traditional competitive factors – the price they receive for their product, the volume of Canadian landings, and the prices of their alternative products (frozen and fresh fillets).[9] In addition, it is assumed that weather conditions make the supply of fish somewhat seasonal. A time trend is added to reflect the symmetric development of supply relationships between independent processors and wholesalers and New England distributors.

The Model for Fresh Fillets

The demand model for Canadian fresh fillets also includes traditional factors, such as the wholesale price of fish, the price of Canadian exports, the volume of U.S. landings, and a measure of seasonal demand. The growth in consumer preferences for fresh fish is captured by a time trend.

The supply model of Canadian fresh-fillet exports assumes that the volume of Canadian landings, the price of fresh fillets, the prices of alternative frozen products, and the time of year influence exporters' decisions. The influence on fillet supply of the set of institutional factors from the industry structure trade model is captured by the inclusion of a time trend intended to account for improvements in quality and freshness and the development of direct sales arrangements between exporters and U.S. supermarkets.

Econometric Estimates

Supply and demand equations are estimated for both whole fish and fresh fillets using a log-linear specification of the equations shown in table 6.1. The whole-fish model also includes an ex-vessel price equation, since ex-vessel prices may be influenced by the quantity of whole-

TABLE 6.1
Industry structure models of Canadian fresh-fish imports

Whole fish

Import demand
WHIMP = C + a_1USIMPRI + a_2USVESPRI + a_3WHPRICE + a_4USLAND + a_5TIME
$\qquad\qquad\qquad\qquad\qquad\qquad\qquad\qquad$ + a_6SPRING + a_7WINTER + e_d.

Import supply
WHIMP = C + a_1CEXPRI + a_2CLAND + a_3FRFILL + a_4FZFILL + a_5BLOCK + a_6TIME
$\qquad\qquad\qquad\qquad\qquad\qquad\qquad\qquad$ + a_7SPRING + a_8WINTER + e_s.

U.S. ex-vessel price
USVESPRI = C + a_1WHIMP + a_2WHPRICE + a_3USLAND + E_p.

Filleted fish

Import demand
FILLIMP = C + a_1WHPRICE + a_2FRFILLIM + a_3USLAND + a_4SPRING + a_5SUMMER
$\qquad\qquad\qquad\qquad\qquad\qquad\qquad\qquad$ + a_6TIME + e_d.

Import supply
FILLIMP = C + a_1CLAND + a_2FRFILL + a_3FZFILL + a_4BLOCK + a_5TIME + a_6SPRING
$\qquad\qquad\qquad\qquad\qquad\qquad\qquad\qquad$ + a_7SUMMER + e_s.

fish exports while at the same time influencing the demand for Canadian whole fish.

A two-stage least-squares procedure is employed to estimate this set of equations, using monthly data for the period 1978–90. This period was selected to reflect a time of sufficient stock abundance for institutional influences to be detected. Because cod and haddock account for the bulk of Canadian groundfish exports to the United States, the data used in the analysis are limited to these species. The definitions of the variables and sources of data are described in table 6.2 (a more detailed discussion of the estimation procedure and the results is provided in the appendix to this chapter).

Table 6.3 reports the estimates of the key parameters in the model. The critical result is that all of the time trend variables are highly significant and positive, indicating that institutional influences have been important. While interpreting time trends as useful proxies for otherwise unmeasurable institutional changes is usually a dubious practice, in this case it is justified because the field research provides independent verification. The econometric findings are wholly consistent with the

TABLE 6.2
Definitions and data sources of variables used in the econometric analysis[a]

WHIMP: Log of quantity of monthly imports of whole cod and haddock from Canada in 000s of pounds (unpublished data of the National Fishery Statistics Program, NMFS, Washington, DC).

USIMPRI: Log of monthly average price per pound in U.S. dollars of imported Canadian cod and haddock from Canada (approximated by U.S. ex-vessel prices of cod and haddock divided by the exchange rate).[b]

USVESPRI: Log of average monthly U.S. ex-vessel prices of cod and haddock (unpublished data of the National Fishery Statistics Program, NMFS, Washington, DC).

WHPRICE: Log of monthly average wholesale price of cod in the United States (NMFS New York Economic Data Office, 'Fulton Fish Market Prices' and daily 'green sheets' of the *New York Market News Report*).[c]

USLAND: Log of monthly U.S. landings of cod and haddock in 000s of pounds (NMFS weighout files).

TIME: Log of trend variable taking the value 1 in January 1978 and increasing by one each month through December 1990.

SPRING: Dummy variable taking the value of 1 for February, March, and April.

WINTER: A dummy variable taking the value of 1 for November, December, and January.

CEXPRI: Log of monthly average price received by Canadian exporters approximated by the U.S. ex-vessel price of fresh whole cod and haddock in Canadian dollars (*Federal Reserve Bulletin* and NMFS weighout files).

CLAND: Log of monthly Canadian landings of cod and haddock in 000s of pounds (DFO, *Canadian Fisheries Landings*, 1978–90).

FRFILL: Log of monthly average U.S. fresh-fillet prices in Canadian dollars (unpublished data of the National Fishery Statistics Program, NMFS, Washington, DC).[d]

FZFILL: Log of monthly average U.S. price per pound of frozen fillets in Canadian dollars (unpublished data of the National Fishery Statistics Program, NMFS, Washington, DC).

BLOCK: Log of monthly average U.S. price per pound of frozen blocks in Canadian dollars (unpublished data of the National Fishery Statistics Program, NMFS, Washington, DC).

FILLIMP: The quantity of monthly fresh fillet imports in 000s of pounds (unpublished data of the National Fishery Statistics Program, NMFS, Washington, DC).

FRFILLIM: The monthly average price of U.S. fresh fillets in Canadian dollars (unpublished data of the National Fishery Statistics Program, NMFS, Washington, DC).

SUMMER: A dummy variable taking the value of 1 for May, June, and July.

---

[a]We acknowledge, with gratitude, Daniel Georgianna and William Hogan for providing us with monthly observations on all the variables for the period 1978–87.

[b]We follow Hogan and Georgianna (1989) by using a proxy for the U.S. import price consisting of the U.S. ex-vessel price of cod and haddock divided by the exchange rate of the Canadian dollar to the U.S. dollar. This step is necessary because of the poor quality of U.S. import price data. Ex-vessel prices were calculated by dividing total landed values by total landings. Monthly landed values and landings for cod and haddock were aggregated for the states of Maine, Massachusetts, Rhode Island, and New Jersey from the weighout files of the National Marine Fisheries Service. The landings from these states accounted for over 90 percent of U.S. cod and haddock landings during the period.

[c]Haddock prices are not always available, and it seems reasonable to assume that the two price series move together.

[d]This was calculated by dividing the total reported value of fresh-fillet imports by the total quantity of imports.

TABLE 6.3
Elasticity estimates of Canadian monthly exports for selected variables: whole and
filleted cod and haddock, 1978–90 (*t*-statistics in parentheses)

Variable	Whole fish Import demand	Whole fish Import supply	Filleted fish Import demand	Filleted fish Import supply
TIME	0.78 (6.55)**	0.49 (4.64)**	0.32 (6.17)**	0.35 (3.86)**
USLAND	−0.38 (−0.91)		−0.35 (−2.1)*	
USIMPRI	−3.79 (−3.60)**			
USVESPRI	3.16 (2.76)**			
CLAND		0.006 (0.058)		0.21 (2.19)*
FRFILL		−2.57 (−4.38)**		2.57 (2.49)*
CEXPRI		1.13 (5.21)**		
FRFILLIM			−0.37 (−1.07)	

* significant at 0.05 level
** significant at 0.01 level

evidence from our field research, and the pattern of the time trend variables in the different regression models corresponds exactly to the predictions of the industrial structure model. For example, the significance of the time trend in the supply of exports corresponds with the findings of our field research regarding the accumulated effect of changes in industrial structure and marketing arrangements, above and beyond the influence of other economic factors on fresh-fish trade. Further confirmation of the importance of industrial structure in the supply of exports can be found by comparing the magnitudes of the time trend parameters in the whole-fish and filleted-fish import supply models.

The industrial structure model argues that the organizational learning in the atomistic sector in southwest Nova Scotia involves discovering how to market whole fish through established distribution channels in New England. The corresponding adjustment in the corporate sector is more complex, involving both establishment of new direct marketing links to U.S. supermarkets and improvement of quality control procedures in order to increase supplies of fresh fillets suitable for sale in supermarkets. Because the adjustments in the corporate sector are relatively more difficult to accomplish than those in the atomistic sector, the corporate adjustment process should be more protracted. This assumption is confirmed by the lower elasticity coefficient on the time variable in the fresh-fillet supply model than is evident in the whole-fish supply model.

In addition, because the corporate-sector vessels must travel long distances to the offshore fishing grounds, while the atomistic sector's inshore vessels can make short trips, the average length of trip is longer and the average freshness of catch is lower in the corporate sector than in the atomistic sector. When stocks are relatively abundant, however, the corporate sector can reduce the length of its trips and significantly narrow the quality gap with the inshore sector.

The differential effect of stock abundance on quality is revealed in these supply equations by the effect of the volume of Canadian landings (a proxy for stock abundance) on exports. As predicted, Canadian landings positively affect exports of fresh fillets, but they do not affect exports of whole fresh fish. This finding is supported by the fact that since 1990 fresh whole-cod exports to the United States have remained constant (except for a dip in 1991) during a period of substantial declines in Canadian stocks, while fresh cod-fillet exports have declined by 61 percent (figure 6.1, table A.15).

The industrial structure model is also supported by structural differences on the demand side of the fillet and whole-fish markets. The time trend variable, for example, partly captures the secular increase in consumer preferences for fresh fish, and this effect should apply equally to both fillets and whole fish. The upward trend in demand in the whole-fish market, however, incorporates the additional effects of initiatives taken by New England processors and distributors to develop new buying relationships with Canadian distributors in southwest Nova Scotia, while our interviews indicated that supermarket chains have been slow to form direct links for purchasing fresh fillets from large Canadian processors.

The slowness with which these institutional changes have developed in the fillet sector means that the effect of time on the demand for fillets should be smaller than on that for whole fish. The econometric analysis confirms that the elasticity of demand for exports with respect to time in the fillet market is, in fact, less than half that of the whole-fish market.

A further structural difference between the demand for exports in the two markets has to do with the sensitivity of export demand to the prices of Canadian fish. In the mass fresh-fillet markets to which the Canadian corporate sector directly markets its product, predictability of supply is more important than price, and this fact is confirmed by the absence of price effects on demand in this market. Because the large distributors and processors in New England switch between New England and Canadian suppliers when relative prices change in the two countries, however, the demand for whole fish from Canada is sensitive to both Canadian and New England prices. This price sensitivity is confirmed by the significant negative effect of rising prices of Canadian whole fish on exports to the United States and by the significant positive effects that increases in New England ex-vessel prices have on demand for Canadian whole fish.

Such simple structural models can only approximate supply and demand behavior in the U.S./Canadian markets for whole fish and fresh fillets. The broad outlines of the model, however, are remarkably consistent with the institutional findings of the field research and are not supportive of models of trade flows that ignore such institutional factors.

SUMMARY

In this chapter the rapidly changing trading relationship in fresh fish between the United States and Canada has been analyzed, focusing on the sharp increases in exports of Canadian fresh whole fish and fillets during the 1980s. Although traditional explanations for changing trade flows shed some light on these rapid increases, they are not totally satisfactory for understanding the bulk of the trade shift.

Exchange rates have not consistently changed in favor of Canadian producers. Likewise, movements in relative prices between frozen and fresh fish in the United States do not correspond to the varying pattern of Canadian fresh-fish exports. Although frozen-fillet prices rose relative to fresh-fish prices, exports of fresh fish increased much faster than frozen-fillet exports. Furthermore, Canadian government subsidies do not appear to have played a significant role in expanding fresh-fish imports. There have been no substantial changes in subsidy levels or utilization rates over the last decade, and the findings of the USITC indicate only a very small influence on Canadian processors' cost advantages from existing federal and provincial subsidies.

The more likely explanations for the increases in imports revolve around the influence of market institutions. This conclusion is supported by our field research and econometric analysis. Canadian government industrial policy has helped to promote substantial concentration in the processing industry. While the highly concentrated and vertically integrated processing sector has traditionally focused on frozen and salt-fish production, recent improvements in processing technology and air transportation networks have begun to open a substantial new segment of the U.S. market to fresh-fish consumption. This new market demands large and reliable supplies of high-quality fresh fish. Although their structure, combined with enterprise quotas, gives large-scale Canadian processors a comparative advantage over U.S. processors in guaranteeing supply, they are at a disadvantage in their inability to produce fish of sufficiently high quality to satisfy growing mass markets for fresh fish, particularly when fisheries stocks are depleted.

Thus, the ability of Canadian companies to expand their exports into the U.S. fresh-fish market hinges on improving the average quality of their product. In contrast, U.S. processors need to offer more predictability of price and supply by acquiring more stable sources of fish. Currently, they are becoming increasingly dependent on Canadian whole-fish exports to meet their supply needs. This supply is likely to diminish, however, as the large Canadian processors improve their networks for selling their products directly to fresh-fish retailers, thus enabling them to circumvent U.S. processors and wholesalers entirely.

Once groundfish stocks in the region begin to recover, all signs point to a long-term increase in Canadian fresh-fish exports to the United States. This trend implies less rapid increases in prices for U.S. fishermen when markets expand, since Canadian exports put a cap on the price increases fishermen are accustomed to during times of reduced landings. In addition, there is likely to be a diminishing role for U.S. processors in brokering and wholesaling fresh-fish products, except in the New England region.

APPENDIX
ECONOMETRIC ANALYSIS

This appendix is intended to provide a fuller discussion of the empirical analysis summarized in the text. The econometric models for the whole- and filleted-fish markets are shown in table 6A.1. Three equations – the demand equation for whole fish (1), the supply equation (2), and the ex-vessel price equation (3) – comprise the model for whole fish, under the assumption that ex-vessel prices, export prices, and export quantities are simultaneously determined. The market for fresh fillets is characterized by a demand equation (4) and a supply equation (5) in which the quantity of exports and export prices are simultaneously determined.

Fresh Whole Fish

In the whole-fish demand equation, we expect $a_1 < 0$, reflecting the first-order demand condition; $a_2 > 0$, since higher U.S. ex-vessel prices should induce processors to substitute Canadian fish for domestic supplies; $a_3 > 0$, since higher wholesale fish prices generate greater demand for imported Canadian fish given the limited quantity of U.S. landings; $a_4 < 0$, because greater U.S. landings reduce the need to supplement supply with Canadian imports; $a_5 > 0$, reflecting the growing adaptation of institutions to the importation of Canadian fish, particularly in terms of marketing relationships (this variable should pick up the increases in imports that cannot be explained by the traditional cost and price variables); $a_6 > 0$, reflecting the findings of our field interviews, that the demand for fresh-fish is highest in this season, partially because of Lent; and $a_7 < 0$, because demand tends to be lower in these months.

For the whole-fish supply equation, we expect $a_1 > 0$, reflecting the usual first-order supply condition; a_2 to be insignificant, since whole-fish exports are assumed to be independent of the size of Canadian landings; $a_3 < 0, a_4 < 0, a_5 < 0$ if higher prices for substitute products lead to a shift in Canadian exports from whole fish to fresh or frozen filleted fish or blocks; $a_6 > 0$, since marketing relationships between Nova Scotia processors and New England buyers have improved over time; $a_7 > 0$, because our interviews revealed that fish caught in this period tend to be fuller and larger and thus are most profitably sold into the whole-fish market; and $a_8 < 0$, since most of the traditional vessel suppliers of whole fish are tied up during this period, owing to rough weather.

The expectation for the ex-vessel price equation is that $a_1 < 0$ if the

TABLE 6A.1
Econometric specifications and expected signs

Whole fish

$$\overset{(-)}{} \quad \overset{(+)}{} \quad \overset{(+)}{} \quad \overset{(-)}{} \quad \overset{(+)}{}$$
$$\text{WHIMP} = C + a_1\text{USIMPRI} + a_2\text{USVESPRI} + a_3\text{WHPRICE} + a_4\text{USLAND} + a_5\text{TIME}$$
$$\overset{(+)}{} \quad \overset{(-)}{}$$
$$+ a_6\text{SPRING} + a_7\text{WINTER} + e_{d}. \qquad (1)$$

$$\overset{(+)}{} \quad \overset{(0)}{} \quad \overset{(-)}{} \quad \overset{(-)}{} \quad \overset{(-)}{} \quad \overset{(+)}{}$$
$$\text{WHIMP} = C + a_1\text{CEXPRI} + a_2\text{CLAND} + a_3\text{FRFILL} + a_4\text{FZFILL} + a_5\text{BLOCK} + a_6\text{TIME}$$
$$\overset{(+)}{} \quad \overset{(-)}{}$$
$$+ a_7\text{SPRING} + a_8\text{WINTER} + e_{s}. \qquad (2)$$

$$\overset{(-)}{} \quad \overset{(+)}{} \quad \overset{(-)}{}$$
$$\text{USVESPRI} = C + a_1\text{WHIMP} + a_2\text{WHPRICE} + a_3\text{USLAND} + e_{p}. \qquad (3)$$

Filleted fish

$$\overset{(0)}{} \quad \overset{(0)}{} \quad \overset{(-)}{} \quad \overset{(+)}{} \quad \overset{(+)}{}$$
$$\text{FILLIMP} = C + a_1\text{WHPRICE} + a_2\text{FRFILLIM} + a_3\text{USLAND} + a_4\text{SPRING} + a_5\text{SUMMER}$$
$$\overset{(+)}{}$$
$$+ a_6\text{TIME} + e_{d}. \qquad (4)$$

$$\overset{(+)}{} \quad \overset{(+)}{} \quad \overset{(-)}{} \quad \overset{(-)}{} \quad \overset{(+)}{}$$
$$\text{FILLIMP} = C + a_1\text{CLAND} + a_2\text{FRFILL} + a_3\text{FZFILL} + a_4\text{BLOCK} + a_5\text{TIME}$$
$$\overset{(+)}{} \quad \overset{(+)}{}$$
$$+ a_6\text{SPRING} + a_7\text{SUMMER} + e_{s} \qquad (5)$$

---

importation of Canadian fish depresses prices for U.S. fishermen; $a_2 > 0$, since higher wholesale prices should increase the demand of processors and wholesalers for U.S. fish; and $a_3 < 0$, since greater U.S. landings depress ex-vessel prices.

Fresh Fillets

In the case of the fresh-fillet demand equation, we expect a_1 and a_2 to be statistically insignificant, since predictability of supply is much more important than are small variations in price in this market; $a_3 < 0$ if increases in U.S. landings have any impact on the demand for Canadian fresh fillets; $a_4 > 0$ because of the seasonally high demand; $a_5 > 0$, since there was some indication in our interviews that the increased demand for fresh fillets extended into the summer season; and $a_6 > 0$, reflecting the growth in consumer preferences for fresh fish.

For the fresh-fillet supply equation, we expect $a_1 > 0$, since higher Canadian landings indicate a greater supply of high-quality fish available to corporate processors for exporting to the fresh-fillet market; $a_2 > 0$ to meet the first-order supply conditions; $a_3 < 0$ and $a_4 < 0$, since they are alternative product forms for fresh-fillet processors; $a_5 > 0$, reflecting the positive influence of changing industrial structure, fishery regulations, and marketing arrangements; $a_6 > 0$ and $a_7 > 0$, since the Canadian catch tends to be higher in these seasons because of weather. (The dummy for summer is not included in the whole-fish supply equation, since it was insignificant and did not substantially change the coefficients or significance of the other variables. Likewise, the winter dummy was insignificant in the fresh-fillet equations.) These equations were estimated using two-stage least squares for the 1978–90 period. In the case of the whole-fish equations, serial correlation was corrected using the Cochrane-Orcutt procedure.

The complete estimation results are reported in table 6A.2. Equation (6) shows that the demand for fresh, whole, Canadian fish is higher, the lower the exchange-rate-adjusted price is, the higher U.S. ex-vessel prices are, and in the spring months (February, March, April). Demand is lower during the winter months (November, December, January). The coefficients on the wholesale price variable and the U.S. landings variable have the right sign but are not statistically significant.

Equation (7) shows that exports of whole fish are larger, the higher the exchange rate-adjusted export price is, the lower the exchange-rate-adjusted prices of fresh fillets and blocks are, and in the spring months. Supply tends to dip in the winter months. The coefficient on frozen-fillet prices is significant but has the wrong sign. We believe this feature to be the result of collinearity among the three price series – fresh fillets and blocks ($r = 0.87$); fresh fillets and frozen fillets ($r = 0.93$); and blocks and frozen fillets ($r = 0.95$). This result is also consistent with the finding that the trend in frozen-fish prices cannot explain the expansion in fresh-fish exports, since frozen prices were rising at a much more rapid rate. As predicted by our model, Canadian landings do not significantly influence exports.

Most importantly, the time trend is significant and positive in both equations, supporting the industrial-structure trade model, which indicates an important role for institutional factors in fresh-fish export decisions.

The price equation (8) reveals that U.S. ex-vessel prices are determined largely by wholesale prices of fresh-fish products and the quantity of U.S. landings.

Equation (9) shows that the demand for fresh filleted Canadian fish is

TABLE 6A.2
Supply and demand estimates of Canadian fresh whole and filleted cod and haddock exports, 1978–90 (t-statistics in parentheses)

Whole fish

Estimated coefficients for demand equation

WHIMP =	C	USIMPRI	USVESPRI	WHPRICE	USLAND	TIME
	6.5	−3.79	3.16	0.51	−0.38	0.78
	(2.45)**	(−3.60)**	(2.76)**	(0.61)	(−0.91)	(6.55)**

				SPRING	WINTER	(6)
				0.35	−0.36	
				(2.92)**	(−2.98)**	

Adj. R^2 = 0.74 F = 55.8   D-W=2.05 N=154

Estimated coefficients for supply equation

WHIMP =	C	CEXPRI	CLAND	FRFILL	FZFILL	BLOCK
	6.37	1.13	0.006	−2.57	3.42	−1.27
	(5.37)**	(5.21)**	(0.06)	(−4.38)**	(4.21)**	(−2.44)*

			TIME	SPRING	WINTER	(7)
			0.49	0.32	−0.31	
			(4.64)**	(3.15)**	(−2.47)**	

Adj. R^2 = 0.76 F = 55.2   D-W = 2.00 N = 154

Estimated coefficients for ex-vessels price equation

USVESPRI =	C	WHIMP	WHPRICE	USLAND	(8)
	2.17	−0.002	0.78	−0.39	
	(5.93)**	(−0.11)	(10.92)**	(−10.82)**	

Adj. R^2 = 0.94 F = 648.6   D-W = 2.18 N = 154

Filleted fish

Estimated coefficients for demand equation

FILLIMP =	C	WHPRICE	FRFILLIM	USLAND	TIME	SPRING
	8.37	0.44	−0.37	−0.35	0.32	0.71
	(4.90)**	(0.94)	(−1.07)	(−2.10)*	(6.17)**	(9.41)**

					SUMMER	(9)
					0.39	
					(4.72)*	

Adj. R^2 = 0.65 F = 47.6   D-W = 2.12 N = 156

Estimated coefficients for supply equation

FILLIMP =	C	CLAND	FRFILL	FZFILL	BLOCK	TIME
	−2.77	0.21	2.57	−1.63	−0.28	0.35
	(2.44)*	(2.19)*	(2.49)**	(−1.45)	(−0.58)	(3.86)**

				SPRING	SUMMER	(10)
				0.70	0.30	
				(7.45)**	(2.40)*	

Adj. R^2 = 0.41 F = 23.6   D-W = 1.72 N = 156

 * significant at 0.05 level
** significant at 0.01 level

higher, the lower U.S. landings are and in the summer and spring months. The statistical insignificance of the coefficients on U.S. wholesale fish prices and the exchange-rate-adjusted price of fresh fillets support our findings that stability and quality of supply, as opposed to small price variations, are the major determinants of supermarket demand for fresh fillets.

The supply equation (10) for fresh fillets shows that exports from Canada are larger, the higher the exchange-rate-adjusted price of fresh fillets is, the larger Canadian landings are, and in the spring and summer months. The coefficients on frozen-fillet and block prices are the right sign but are insignificant. The exchange-rate-adjusted export price of whole fish was not included in the equation, since our interviews revealed that whole-fish exports are not a significant alternative product for exporters of fresh fillets. Again, the time trend variable is positive and significant, supporting the possible role of institutional factors in export supply decisions.

To test the robustness of our specifications over different time periods, we estimated the equations for the 1978–86, 1978–87, 1978–88, and 1978–89 periods. The equations remained substantially unchanged. The only change in the fillet equations was that the coefficient on U.S. landings in the demand equation was insignificant in the two earliest periods, and the coefficient on Canadian landings in the supply equation was insignificant for two of the middle periods. The whole-fish, ex-vessel price equation remained unchanged, while the winter dummy was insignificant in the early period supply equations, and the coefficient on the wholesale price was significant in the 1978–86 demand equation.

7

Conclusions for Policy

It has been shown in the preceding chapters how the fishing industries in New England and Atlantic Canada have evolved, how their private and public policy institutions have interacted with markets, and how these institutions translate into differences in competitive advantage. As Canada has expanded its presence in U.S. fresh-fish markets, however, a variety of trade, boundary, and regulatory conflicts has emerged that are linked to the problems inherent in managing a common-property resource. These developments are placing new demands on industry structure and public policy. Following a brief review of the major findings of the study, the policy choices facing the two countries are explored in this chapter.

OUTPUT AND REVENUE

Trends in output and revenue in the fishing industry are subject to various factors beyond the control of individual fishermen, processors, and wholesalers. Output is naturally volatile because of biological shocks to the resource – weather and climatic conditions, water temperatures, and the vagaries of disease and predation. Overlaying this natural uncertainty in production are the effects of common-property incentives, government regulatory and industrial policies, changing dietary preferences, and pure luck. Volatile output translates into volatile revenue, particularly in the spot markets for fresh fish. Although these factors make year-to-year comparisons of the economic condition of the industry difficult, certain major trends over the last two decades can be identified.

During the late 1960s and early 1970s groundfish and scallop harvests dropped precipitously in both countries as large-scale factory trawlers from western Europe and the Soviet Union intensively fished the offshore waters of the northwest Atlantic. After Canada and the United States extended their coastal boundaries to 200 miles under the provisions of the Law of the Sea Conference in 1976, domestic output rebounded strongly. The catch of key groundfish species rose by 50 percent between 1977 and 1980 in New England and by 55 percent during the same period in Canada (table A.2).

Over the next few years increased U.S. fishing efforts cut deeply into stocks, and groundfish landings generally declined after 1980. The more conservatively managed Canadian stocks continued to increase until 1982, and since then there have been occasional good years in the harvest of specific Canadian stocks. The core groundfish stocks have declined precipitously in the 1990s, however, leading to closures of fishing grounds in Atlantic Canada and the imposition of unusually strong regulatory restrictions in New England.

Declines in the New England catch in the 1980s were offset by increases in prices as demand for fresh fish increased. Although there was considerable variation in year-to-year values, New England catch value held up in real terms through the catch declines of the mid-1980s. The real value of catch in New England exceeded its 1980 value in 1983 and 1984 and was only slightly below the 1980 value by 1985. In Canada real values declined until 1985, when price increases and a shift in the composition of exports to the United States led to a dramatic rise in the real value of catch. By 1987 the real value of Atlantic Canada landings was 43 percent above that of 1980. By the late 1980s, however, the combination of falling landings and slower increases in fish prices resulted in declining real value of revenues.

Growth in employment in the industry tended to parallel growth in landings and revenue. For example, the large increase in landings and revenue after the 200-mile limit was established brought an influx of new harvesting and processing workers into the industry. Employment responded more slowly to the reductions in catch and revenue that followed. The work-sharing practices on kinship vessels and in paternalistic processing plants, coupled with unemployment insurance payments and the lack of alternative job opportunities, contributed to 'sticky' labor and lagged employment adjustments in many parts of the industry.

CHANGES IN PRODUCT MARKETS

Both the United States and Canada harvest the same major groundfish species – cod, haddock, flounder, pollock, hake, whiting, and ocean perch – from contiguous resource pools in the northwest Atlantic. His-torically, the industries in both countries served primarily domestic markets for fresh, salted, and frozen fish. Over time, however, the fro-zen- and salt-fish industries became globally competitive commodity markets, while the fresh sector remained relatively specialized and con-tinued to be confined largely to domestic markets. As these markets changed, Atlantic Canada (with its larger stocks and smaller domestic market) concentrated more on the salt and frozen markets, whereas New England producers have for many years sold almost exclusively into the fresh-fish market.

In the 1980s the Canadian industry sought to move production into higher-valued products. Doing so involved improving the quality of fresh and frozen production and expanding the Canadian share of the growing U.S. market for fresh fish. The result has been increasing inte-gration of markets in the two countries – Canadian fresh fish are now sold in large amounts in the expanding and geographically diversified U.S. market, both countries rely on many of the same processors and distributors for marketing fresh fish at the wholesale level, and Cana-dian frozen-fish processors now have branch plants in the United States.

CHANGES IN INDUSTRIAL STRUCTURE

The industries in both countries started with similar industrial struc-tures, common technologies, and little vertical integration. As each coun-try began to specialize in different product lines, however, it adopted markedly different institutional strategies and industrial structures.

The industrial structure of the New England fresh-fish industry remains much as it has been for decades. Fishermen are independent owner-operators, processors are relatively small, and there are numerous wholesalers and distributors that broker fish among processors and sell to wholesale and retail outlets. The atomistic structure of the New England system does not permit significant economies of scale, and there is no formal vertical integration between harvesting and processing.

In contrast, Canada has developed a large-scale, vertically integrated corporate sector that owns a captive fleet. This sector has long domi-nated frozen production and much of fresh-fish processing as well. There has been a trend in Atlantic Canada toward increasing concentration in this vertically integrated sector through mergers and closings of processing firms and increased concentration has meant expanded access to scale economies in harvesting, processing, and distribution.

Atlantic Canada has also retained an atomistic sector of independent fishermen and processors similar to that of New England. This sector produces for salt-fish and occasionally for frozen-fish markets, and it may also harvest shellfish, such as lobster, but its most prosperous market is for whole fresh fish that are exported to New England.

LABOR MARKET STRUCTURE

The industry has a unique set of labor market institutions that vary across its different sectors and have consequences for efficiency and income distribution. There is a kinship sector in harvesting that emphasizes work and income sharing and family obligations that motivate work effort, and there are also independent processors that rely on paternalistic employment relationships to recruit, retain, and motivate a highly seasonal labor force. The corporate capitalism sector, consisting of some of the offshore vessels in the New England fleet and the vertically integrated corporate sector in Canada, uses high wages and efficient management practices to maintain labor productivity.

Each of these employment systems responds differently to economic change. The kinship and paternalistic systems share out available work among the existing labor force in ways that retard employment adjustments and contribute to high levels of labor force attachment to the industry. In contrast to the 'sticky' labor of the kinship and paternalistic capitalism sectors, the corporate capitalism system emphasizes highly variable labor inputs and 'lean' employment levels.

There are also differences in unionization and collective bargaining across these three types of employment systems. Kinship and paternalistic employment practices, combined with an atomistic industrial structure, have inhibited unions. Conversely, unions are more prevalent in the corporate capitalism sector, particularly in Atlantic Canada, where unions have had an increasing influence over wage setting and work rules over the past two decades.

PUBLIC POLICY INFLUENCES

The divergence in industrial structures, labor market institutions, and market focus between the two countries has not been solely the result of competitive business strategies. Public policy has played an increasingly important role in shaping industry differences in both Canada and the United States.

Both countries have addressed common-property problems by regulating catch. In the United States, a federal-state system provides a major role for industry representatives in setting management policies. This system gives the fishing industry considerable latitude for self-regulation and is based on achieving consensus among competing interest groups in the industry. The result has been unrestricted entry and, although catch quotas were tried briefly, a reliance upon a complicated set of area closures and gear restrictions as the principal tools for limiting catch. Conflicts among economic interest groups, long delays in the development of plans that are acceptable to the federal government, and overharvesting pressures have placed the U.S. regulatory process in a state of perpetual crisis.

In Canada, management is also a federal government responsibility, but the balance of power lies with policy-makers in the federal government rather than with industry interest groups. The Canadian regulatory process has used various licensing mechanisms to restrict entry and has adopted a mixture of transferable quota allocations (applied to large processors and recently to parts of the inshore fleet) and overall catch quotas (applied to some groups of smaller independent vessels). This combination of insulation from economic interest groups and utilization of more readily monitored regulatory regimes has led to relatively more effective regulation of stocks in Atlantic Canada than in New England.

Regulatory policy also controls the distribution of catch between large-scale and small-scale fishermen and processors, as well as the overall level of output. In some years, the regulatory process appears to have favored independent fishermen, while in other years the larger processing corporations have benefited. These shifts in catch shares, however, have been relatively small.

The instability of catch and prices, coupled with the fact that fishing activity is concentrated in isolated rural ports and in relatively weak regional economies, also has made the Canadian fishing industry a target for industrial and regional development policy. In the area of industrial policy, both countries have adopted capital subsidy and/or fleet modernization schemes, provided monies for port and harbor development, and used unemployment insurance benefits to augment the seasonal incomes of fishermen and processing workers. The greater dependence of Atlantic Canada on the fishing industry, however, has meant that such policies have been more extensive in Canada than in the United States.

In addition, Canada has also used regulatory policy as an economic development tool. Both licenses and quotas in Canada are allotted on a geographical basis, so that regulatory policy can determine the regional locus of economic activity.

The most notable policy intervention has been in Atlantic Canada, where the government directly altered the structure of the industry. In 1984 the federal government engineered the merger of five of the larger processing companies into two vertically integrated corporations – a step deliberately intended to strengthen the competitive position of the industry. Part of the merger package involved substantial infusions of public capital. In contrast, the industry structure in New England has been largely free of government intervention.

COMPETITIVENESS AND TRADE

Changing business and labor institutions and the influence of public policy have affected competitiveness and trade patterns between the two countries. Atlantic Canada exports 80 percent of its catch, three-fourths of which is sold to the United States (primarily through New England markets). This trade has traditionally been in frozen product lines, but there has been increasing head-to-head competition between New England and Atlantic Canada in U.S. markets for fresh fish. A growing export capacity and falling New England landings have shifted market share in favor of Canadian production. Canadian fresh whole-fish imports, for example, rose from 4 percent of U.S. landings in 1979 to almost 30 percent in 1985, and similar increases have occurred in imports of fresh processed fish. Many of these increases have been concentrated in the late winter and early spring months when fresh fish command the highest prices.

This trade surge has been explained in various ways. Economists, for example, look to changes in factor costs and exchange rates as the key factors that determine trade flows, while the U.S. industry attributes it to government subsidies and the transfer of fishing stocks from the United States to Canada following the World Court boundary determination in 1984. None of these explanations, however, is consistent with the timing of the trade surge or with estimates of the effects of prices and subsidies on trade flows. In addition, all of them underestimate the importance of strategic responses by business and of the integration of fisheries regulation with the structure of the industry. Instead, the changing pattern of fisheries trade underscores the importance of interactions among institutions and markets.

Dual Production Systems

Differences in the business, labor, and public policy institutions in the two countries, operating within common market and technological contexts, have led to two different production systems. One is the atomistic and relatively loosely regulated system of independent fishermen and processors found in both countries. This system is guided largely by market competition. The other is the large-scale vertically integrated system, unique to Atlantic Canada, that is governed by oligopolistic competition and discretionary economic practices. Each of these systems has contributed in different ways to the increase in Canadian exports of fresh fish to the United States.

Exports from both Canadian production systems were stimulated in the late 1970s by an abundant fisheries resource base and relatively conservative management practices, compared with those of the United States. This standard factor cost explanation was reinforced by rising fresh-fish prices in U.S. markets and, along with improved marketing links between the two countries, it accounts for much of the increase in whole-fish exports from Canada to the United States.

The vertically integrated production system in Atlantic Canada has further benefited from a series of institutional factors that were not available to the atomistic sector. One factor is the enterprise quota system, which allows the vertically integrated corporate sector to target harvesting effort by region and season and to direct landings to the most efficient processing facilities. The presence of enterprise quotas allows scale economies to be fine tuned around the most efficient plants. Moreover, the quotas have allowed fresh production to be concentrated in the late winter and early spring months, when smaller vessels are unable to fish and New England fresh-fish prices are at their peak. Although these advantages have helped large Canadian processors to compete with the independent sector for traditional New England markets, their greater importance lies in mass markets where high-volume fresh-fillet orders can be directly negotiated with U.S. supermarkets.

Mass markets for fresh fish differ substantially from both frozen-fish markets and traditional white-tablecloth markets for fresh fish. Mass markets demand fish of high quality to ensure long shelf life and, at the same time, require guaranteed future supplies at predictable prices. The institutions of large-scale, vertically integrated production, combined with the enterprise allocation scheme for regulating catch volumes in the corporate sector, are well suited both to serve mass markets and to realize the advantages of scale economies in marketing and distribution that are possible in such markets.

Although the large-scale, vertically integrated Canadian producers have the ability to commit to high-volume sales at a predetermined price, they have been traditionally more attuned to the quality standards of frozen, rather than fresh, fish markets, so that quality has been a barrier to shifting production more rapidly from frozen to fresh fish. Nonetheless, during the early 1980s, when fish stocks were relatively abundant, the corporate sector began to enter the market for fresh fish in New England and to sell to supermarkets in midwestern states, where mass markets for fresh fish have historically been underdeveloped. These sales have been an important factor in stimulating the surge of fresh-fillet exports to the United States. Growth rates in the fresh-fillet market, however, have been limited by the inability of the corporate sector to secure larger volumes of high-quality fresh fish and by the cost of creating marketing channels in the midwest.

In contrast to the corporate sector, the atomistic sectors in New England and Atlantic Canada have a long tradition of quality. New England fishermen have been the main suppliers of white-tablecloth markets and supermarkets in the northeast, whose customers have traditionally accepted unpredictable availability of specific species and fluctuating prices. In the early 1980s whole fish from Nova Scotia's independent sector began to penetrate this market as prices in New England rose.

Because the atomistic sector in Atlantic Canada has relied upon many of the same distribution channels used by New England fishermen, the atomistic sector has found it easier than the corporate sector to penetrate U.S. fresh-fish markets. The lack of enterprise quotas and fragmented production, however, make it more difficult for this sector to ensure the large-scale deliveries at relatively predictable prices required for competing against the Canadian corporate sector in mass markets.

FROM POLITICAL ECONOMY TO PUBLIC POLICY

The mix among production systems affects the efficient level of catch, and the efficiency of particular systems is intimately connected to the specific regulatory instruments used to achieve efficient catch levels. In effect, the question of the efficient balance among different production systems cannot be separated from regulatory and other policies and therefore from policy decisions regarding the distribution of income within the fishing industry. In this section the policy choices facing the United States and Canada are explored.

Regulating Catch Levels

Despite common regulatory goals and the recent stock collapses in both countries, regulatory targets are generally believed to have been set closer to efficient levels in Atlantic Canada. Moreover, Canadian regulatory practices are thought to permit relatively better enforcement of harvesting targets than those in the United States. The Canadian regulatory approach has, in principle, many advantages in achieving conservation objectives and in directing economic activity. It can deal directly with capacity by limiting entry of vessels and fishermen, and the enterprise allocation approach can directly limit fishing effort by those vessels with the largest capacity. The regulatory arrangements governing parts of the independent sector in Atlantic Canada, however, are subject to many of the same problems that have led to overfishing by the New England fleet, and the recent collapse of Canadian stocks is partly attributable to difficulties in enforcing aggregate catch targets on the atomistic, independent fleet. This enforcement problem has led the Canadian government to extend transferable quotas to a significant portion of the Nova Scotia inshore fleet.

In both countries, the stock collapse is the immediate problem to be overcome. While faulty biological assumptions and environmental factors have contributed to this problem, pressures from independent harvesters are a root cause, which can be addressed only by direct restrictions on effort and capacity in this sector. They include tighter restrictions on entry, individual or enterprise quota limits on larger vessels, and individual quotas, taxes, or well-enforced limits on catch by small vessels.

Regulation and Diversity in Production Systems

A second, and little-recognized, dimension of the regulatory problem is the importance of production flexibility. Differences in the flexibility and adaptability of different production systems need to be recognized if efficient resource utilization is to be achieved. For example, some systems respond more readily to variability in allowable catch levels. Smaller independent vessels in both countries are well suited to adapting to year-to-year and even month-to-month changes in catch levels. Such vessels have flexible gear, can readily shift among groundfish species and lobsters as biological availability changes, and have established sales channels in markets that can accommodate changes in the mix of species harvested.

In contrast, gear changes are more costly on large offshore vessels. Moreover, the corporate capitalism sector in Atlantic Canada is tied to mass markets for fresh and frozen fish. This combination of mass production and mass distribution makes it more difficult for harvesting operations in the corporate sector to accommodate short-term changes in the mix of species harvested.

Although these differences in responsiveness have been used simply to advocate relaxed catch targets for the small-scale sector, a more important point is the recognition that regulation must be tailored to the efficiency characteristics of each production system. The large offshore-vessel system needs catch allocations that are relatively stable from year to year, particularly where harvesting is tied to the mass production and marketing of fish, while the flexibility of small-scale, independent vessels supports variable catch allocations to that sector. Concentrating the consequences of biological and regulatory instability in the small-scale sector for efficiency reasons, however, is likely to meet with political opposition from independent fishermen and processors during periods of declining catch allocations.

A second consideration involving regulation with diverse production systems is the extent to which factor inputs into harvesting and processing are fixed. In the case of labor inputs, for example, the corporate capitalism sector can quickly adjust its workforce to changes in catch levels. Under paternalistic capitalism, the number of employees also varies with the level of catch, but work rotation practices designed to spread eligibility for unemployment insurance throughout the processing workforce are used to maintain an inefficiently large, peak-load labor reserve. The attachment of surplus labor to the industry is even stronger under kinship capitalism, because family employment obligations preclude downward adjustments in employment when catch declines.

Where labor is variable, as it is under corporate capitalism, there is a single margin of adjustment defined by factors such as seniority and the marginal productivity of labor in the industry. Where labor inputs are more or less fixed, as they are under paternalistic and kinship capitalism, adjustment occurs at a different rate and along different margins, so that the least efficient labor is not necessarily the marginal labor in the industry.

The presence of 'sticky' labor has implications for regulation. On the one hand, the combination of sticky labor and employment adjustments that are not based upon the marginal productivity of labor means that the social opportunity cost of labor employed in the industry is lower than would otherwise be the case. As social opportunity costs fall, the efficient level of catch for regulatory purposes increases (Terkla, Doeringer, and Moss 1988). Moreover, as a result of lags in labor force adjustment caused by sticky labor, year-to-year variations in regulated catch levels will be more efficient if they follow a lag structure that corresponds to that of each employment system.

A further implication of labor attachment is that labor mobility responds asymmetrically to fluctuations in output. Rising revenue draws labor into the industry relatively quickly, but labor recruited to the kinship and paternalism sectors is not readily released. It is important, therefore, that regulatory policies reflect the need to control the entry of labor, as well as capital, into the industry to constrain harvesting capacity to efficient levels.

At present, these considerations are largely ignored by the regulatory process in both countries. Catch levels and regulatory practices in New England are set without regard to differences in production processes, type of market, or type of employment system. In Atlantic Canada, regulatory distinctions are drawn between the corporate capitalism sector and the kinship sector, but neither the Canadian estimates of optimal catch levels nor the allocation of catch between sectors reflect differences in adjustment flexibility.

Proposals have been made in some fisheries to decentralize the regulatory process to the community level to take advantage of social forms of regulatory control (Ruddle, Hviding, and Johannes 1992). Such decentralization would allow for the adjustments in catch levels needed to incorporate the efficiency implications of sticky labor. Even a decentralized regulatory system, however, would not resolve problems of overfishing by vessels not subject to community-based social controls, such as capitalist vessels and vessels from different communities that are sharing common fishing grounds.

Harmonization of Regulation and Economic Development

Institutional sources of labor stickiness are reinforced by the underdevelopment of many fishing communities and by unemployment insurance practices, particularly in Atlantic Canada. If alternative employment prospects for fishing industry labor were to be improved, there would be increased incentives for surplus labor to leave the industry, the social opportunity costs of fishing labor would increase, the amount of sticky labor would be reduced, and optimal catch levels could be adjusted accordingly.

Although the New England economy has achieved a higher level of income than that of Atlantic Canada, neither country has been able to make much progress in diversifying its weakest port communities in order to reduce the problem of labor stickiness. Lack of effective economic diversification has led to a number of alternative policies – fisheries industry subsidies, income transfer programs, and (in Atlantic Canada) regulatory allocations of catch – which have had the effect of further reinforcing the attachment of labor to the industry and have made regulatory reform more difficult.

The most constructive approach to achieving greater adjustment efficiency would be to decouple more fully these policy instruments from the fishing industry. Industry subsidies would be treated as part of general development incentives, rather than being dedicated to the fishing industry; special fisheries unemployment insurance schemes would be integrated with regular unemployment insurance; and structural adjustment programs of training and relocation would be linked to jobs outside fishing.

Labor Market Adjustment Policies

While economic development policy is one means of addressing the problem of sticky labor in the fishing industry, it must be accompanied by adjustment programs that facilitate the movement of labor to other industries. Labor market information, retraining, and relocation programs are the common components of such adjustment policies. The effectiveness of these policies is often limited by the characteristics and job preferences of the fisheries industry workforce. Particularly for harvesters, earnings during periods of prosperity are considerably higher than those received by comparable workers in other sectors (Doeringer, Moss, and Terkla 1986a). The expectation of periodically high earnings, when coupled with the nonpecuniary attractions of fishing as a way of life and the economic security of kinship and paternalistic employment practices, means that alternative employment must pay relatively high earnings to induce sticky labor to leave the fishing industry. The higher the wage needed in alternative employment is, the higher are the costs of economic development incentives and of human-resources adjustment programs.

Given these cost impediments, economic development and labor market adjustment policies are unlikely to provide sufficient reemployment opportunities for the surplus labor in the industry, particularly in the short term. Where there are not enough jobs, training programs can become income-maintenance programs, and pressures for direct income subsidies increase. This scenario is already apparent in short-term responses, in both Atlantic Canada and New England, to the stock collapse. The restrictions on fishing effort following the stock collapses in both countries prompted various emergency measures – industry and community development assistance, retraining programs, and expanded unemployment insurance – to alleviate financial hardship and to develop employment alternatives for the fishing industry in the short term. Many of these measures, however, remain tied more to the preservation of the fishing industry and its workforce than to reducing capacity sharply. Only recently have policies that reduce harvesting capacity, such as vessel buyout programs, been seriously considered.

Labor market adjustment needs to be predicated on helping to shape a future in which the industry has less harvesting and processing capacity and a much smaller workforce than it did in the late 1980s. Adjustment programs to reach this leaner industry should recognize both the immediate hardships on the workforce and the importance of containing employment to a level that is consistent with efficient catch targets. The latter will be especially difficult because of the economic incentives for entry that are inherent in efficiently regulated common-property industries.

For those younger workers who will be permanently in surplus, economic development and retraining programs are the preferred policy option, while income maintenance and early retirement may be more appropriate policies for more senior workers. Income maintenance is also the most straightforward way of alleviating the current hardship for that fraction of the workforce that will be reemployed when stocks rebound.

One challenge for these policy goals is to find a workable means of distinguishing between those who can reasonably expect to be part of the permanent fisheries workforce and those who cannot. A second is to avoid having income maintenance reduce adjustment incentives for those who should seek alternative employment, or to contribute to unnecessary labor stickiness in the industry.

Policies toward Industrial Structure

The atomistic and loosely regulated industries in New England and southwest Nova Scotia are structured appropriately for harvesting a variable supply of fish that can be priced flexibly and sold fresh to white-tablecloth markets and to those supermarkets where unpredictability of supply and price is tolerated by customers. Although this market is growing, it has been unlikely to become sufficiently large to absorb the catch from this sector, assuming stocks are rebuilt.

In contrast, the vertically integrated, corporate capitalism sector in Atlantic Canada is ideally suited to serve mass markets for frozen and fresh fish, particularly when stocks are rebuilt and catch allocations stabilized. In the longer term, high-value mass markets for fresh fish are where the potential lies for rapid market growth. This market is likely to be dominated by Canada unless there is either a restructuring of the New England industry along Canadian lines, or the independent sector in both countries develops business structures and alliances with mass marketers that can compete with vertically integrated production in mass markets for fresh fish in the midwestern United States.

Without public policies to encourage restructuring, the prospects are poor for reorganization, by either the New England industry or the independent sector in Canada, in order to compete in these midwestern mass markets for fresh fish. Even the large-scale Canadian industry, which had a first-mover advantage in entering mass markets for fresh fish because it was already structured to produce and market efficiently for highly competitive frozen-fish markets, has had problems in structuring production and achieving the quality levels necessary for mass marketing fresh fish in the United States.

Yet neither the atomistic Canadian industry nor the New England industry has any experience with vertical integration and large-scale production. While New England does have a mass production frozen sector with distribution networks to supermarkets, this sector remains an 'enclave' industry that acquires its fish from global markets, and it has shown no inclination to enter fresh-fish markets. Similarly, even though there is some informal vertical integration between processors and the fishermen who regularly supply them, and there are large distributors that pool unpredictable supplies of fish from throughout New

England, fragmented production coupled with the absence of a regulatory process that stabilizes landings constitute a barrier to producing the predictable supplies of fish at predictable prices that are the prerequisites for opening new mass markets.

Price Determination and Quality Incentives

It is in the interests of both countries to shift their production toward the highest value-added markets. The New England industry has accomplished this change largely by concentrating on fresh fish and by dominating white-tablecloth markets that demand the freshest and highest quality catch. Canada, on the other hand, still sells substantial volumes of fish into frozen and salt-fish markets that can tolerate lower quality. The Canadian industry has been trying to move into more upscale markets, but it has experienced difficulties in achieving quality improvement. While some aspects of quality, such as freshness, will always be limited by Canada's greater distance from U.S. markets, changes in the length of fishing trips, in the handling of catch on board vessels, and in processing can improve quality.

Prices are the main incentive for quality improvement and are set under a variety of institutional arrangements in the two countries. New England now has three auction markets, each with its own set of market rules. Although these auctions provide 'reference prices' for the entire region, by far the largest number of transactions occur through bilateral negotiations between fishermen and processors or distributors.

Apart from Portland, where there is a 'display' auction, catch is bought sight unseen in New England markets (and the entire vessel catch is auctioned as a block in New Bedford), so that there is no opportunity for quality differences to be accurately reflected in selling prices. When there are bilateral negotiations over the final amount to be paid for catch value, the ambiguity of quality and the unequal bargaining power between the relatively small group of onshore processors and the more numerous small fishermen are thought to depress prices.

In Canada, there are no auction markets, and prices among independent fishermen and processors are determined bilaterally. In the corporate sector, fish prices are set administratively through internal accounting procedures or through bilateral negotiations with independent fishermen and their associations. Fresh fish harvested by the independent sector that are trucked to New England are subject to the same reference prices as the New England catch. Apart from the occasional cooperative, small-vessel fishermen in southwest Nova Scotia have few market alternatives to selling to the corporate sector in Atlantic Canada or to distributors in New England.

Despite frequent complaints about product quality, allegations of collusion and price fixing, and concerns about information imperfections, existing pricing and marketing arrangements should not be seen as a major source of policy concern. For one thing, the rising demand for high-quality fresh fish should create pressures for more accurate valuation of product quality. Second, the presence of independent, mid-size processors in Canada is providing an additional offset to the market dominance of corporate processors, and the expansion of exports of fresh fish to the United States by independent distributors is generating more competition among processors in both Atlantic Canada and New England.

Although many New England fishermen have a choice among market outlets – traditional auctions in New Bedford and Boston, the display auction in Portland, and direct negotiations with processors – there is little evidence that one market institution is coming to dominate the others over time. Most fishermen seem to exhibit stable preferences for either auction sales or bilateral negotiations. The absence of significant switching among alternative market institutions is characteristic more of a competitive equilibrium than of markets with major distortions from bad information or collusion.

Perhaps the best example of this choice-based test of market efficiency is the experience of the new Portland display auction. The Portland auction market, because it permits greater quality differentiation of fish and because it eliminates bilateral post-auction negotiations, should attract fishermen away from less advantageous selling arrangements and force buyers into open competition for catch. Although Portland is becoming established as a regional market for northern New England fishermen and has attracted a core of buyers and fishermen, there is little evidence of the kind of massive switching of buyers or sellers that would indicate substantial failures in other markets.

Moreover, whatever information problems and quality disincentives exist in bilaterally negotiated prices, these factors are less likely to affect the corporate-sector, where quality standards, trip lengths, and internal catch prices are set administratively. The corporate sector has been instituting steps to improve quality grading and to build quality incentives into its wage payment system. New arrangements for supplementing quality landings by the captive fleet with supplies from independent fishermen are also being tested. Nevertheless, interviews with corporate-sector managers suggest that these quality-improvement measures require considerable time to perfect and that they are dependent on the restoration of well-managed stocks.

POLICIES FOR COMPETITIVENESS AND TRADE

New England fishermen have repeatedly claimed that subsidies to the Canadian industry have resulted in an unfair competitive advantage for Canadian fishermen. Despite the frequency of hearings by the International Trade Commission, there has been only a single finding (in 1986) of an unfair trade advantage, and the level of the countervailing tariff imposed in that case suggests that subsidies have had only a small effect.

The frequency of complaints to the USITC, however, is symptomatic of the concern among New England groundfishermen over the competitive potential of the Canadian industry and its adverse consequences for their economic interests. The large Canadian biomass and conservative regulatory policies have been a more important source of competitive advantage than have subsidies. Enterprise quotas and large-scale vertical integrated producers are likely to reinforce this advantage, particularly once Canadian stocks are rebuilt and the corporate sector learns how to develop mass markets in the United States.

Trade restraint, therefore, is a dead-end policy, which diverts attention from more fundamental differences in competitiveness in Canada and the United States. What is needed instead are more efficient industrial structures, harmonized regulatory policies, and a focus on policies that support the integration of markets in the two countries.

Institutional Reform and Trade

The fresh-fish industry in the United States has been the growth market for high-value fisheries products. This market has an even stronger growth potential if mass markets for fresh fish in the midwest can be developed. As New England landings have fallen, Canadian production has increasingly penetrated the existing fresh-fish market; but enlarging the U.S. market has proved more difficult.

The critical foundation for market expansion is a large and reliable supply of very high-quality fish. The current collapse of stocks in both countries has temporarily disrupted this supply, and rebuilding these stocks is the sine qua non for market growth. Nonetheless, even when landings were higher, each country lacked a critical institutional element needed to position it effectively to serve potential growth markets. Now that free trade has been established, and assuming that stocks will be rebuilt, overcoming these institutional limitations in each country remains essential to achieving growth in the market for fresh fish.

Prosperity for the U.S. industry will depend upon institutional adaptations in industry structure and regulatory policy. The industry must be allowed to deliver reliable supplies of fish needed to open new markets. Canadian industrial structure and enterprise-based regulatory policy are more suited to serving these new markets. However, the capacity to provide a high-quality product needs to be improved.

Such institutional changes can assume a number of different forms. The Canadian industry can develop the quality control needed to sell volume product to U.S. food retailers and then bypass the U.S. processing and distribution system entirely by selling directly to supermarkets. Alternatively, it can develop formal alliances with the U.S. marketing network and rely on this enhanced network to open new markets. Likewise, the United States can elect to change dramatically its regulatory policies to more closely resemble those of Atlantic Canada, and the industry can then adopt an industrial structure similar to that of the large-scale Canadian processors. Failure to make such adaptations will result in both countries' funneling their product into existing markets, thereby missing a significant opportunity for market expansion.

Transboundary Partnerships

Regardless of the extent of institutional change, both countries can benefit from a greater integration of the northwest Atlantic regional fishing economy. Even at present stock levels, Canada has sufficient biomass potential to meet the foreseeable U.S. demand for fresh fish. New England processors and distributors have much to offer the Canadian industry in terms of quality control, marketing contacts, and distribution networks. If the New England industry cannot soon reach some accommodation with the Canadian industry, however, this advantage will be lost, since Canada will develop its own capabilities in these areas.

This regional integration of the fishing economy should also be extended to include reciprocal harvesting and landing rights and the encouragement of transboundary capital and labor mobility. For example, under current statutory and boundary arrangements, Canadian and

American vessels are barred from fishing in each other's territory and Canadian vessels cannot land catch in American ports. These arbitrary demarcations, within the context of common resource pools and common markets, breed inefficiencies for fishermen, processors, and consumers in both countries, since trawler tows must be aborted when vessels approach the boundary line, and Canadian fish must be transported over unnecessarily lengthy overland routes to arrive in American markets.

Integration in harvesting should also be coupled with harmonized management practices, particularly in contiguous fishing zones, such as those on Georges Bank. One market, with a common resource pool, is not well served by two distinct regulatory arrangements. Although the regulatory legislation in both countries has remarkably similar goals, the differences in regulatory instruments, regulatory procedures, and standards for catch levels have been serious obstacles to joint management. This situation is largely the result of three factors – the importance of regulatory policy as an element of regional development policy in Canada, the incompatibility of Canadian-style enterprise allocations with the atomistic and independent harvesting industry in New England, and the differences in the distribution of power among the industry interests as they are vested in the regulatory process in both countries. Developing joint regulatory plans would eliminate one major obstacle to cross-boundary fishing. Removing the prohibition on Canadian vessels landing catch in American ports would remove a second obstacle.

There are also restrictions on the entry of U.S. processing firms into Canada as a result of the preference given to Canadian firms in the enterprise allocation formulas. There are similar restrictions on labor migration from Atlantic Canada to New England. The regional integration of politically troublesome industries like fishing is most easily accomplished within a framework of more general economic integration.

INSTITUTIONAL VERSUS MARKET FAILURES

The principal theme of this book is that institutional, as well as market, factors can affect economic performance and that there can be institutional as well as market failures. The various institutions affecting the fishing industry in New England and Atlantic Canada boldly illustrate this point. Industrial structures and regulatory practices have been often out of sync with changing markets on both sides of the border.

Changes in institutions are much slower and more erratic than changes in markets. Needed restructuring of the Canadian industry was progressing too slowly for the industry to remain competitive until the pace of change was accelerated by government intervention. Regulatory policy in the United States has been beset by sufficient conflict to prevent significant preservation of the stocks of key species. Improvements in quality have been hard to achieve in the large-scale processing sector in Canada, and independent fishermen and producers have failed to devise institutional solutions that would gain them entry into emerging mass markets for filleted fish in the United States.

Institutional failures, unlike many market failures, are inherently transitory. But the path-dependent character of institutional adjustment, the political pressures to protect fisheries jobs and incomes, and the consequent lags in the adjustment process compound to produce long delays in institutional change. The political economy of the north Atlantic fishery clearly illustrates the limitations of market forces in ensuring the institutional changes required for industrial competitiveness and highlights the need for public policies to rectify institutional, as well as market, failures.

Appendix

TABLE A.1
Landings, all species, Atlantic Canada and New England: 1977–93

	Atlantic Canada	New England
1977	1,003,074	263,650
1978	1,153,231	299,697
1979	1,237,702	321,419
1980	1,156,088	357,472
1981	1,194,557	316,153
1982	1,197,632	311,774
1983	1,108,439	322,539
1984	1,065,205	314,930
1985	1,187,937	267,525
1986	1,244,957	251,589
1987	1,266,486	247,275
1988	1,376,149	258,505
1989	1,284,036	256,285
1990	1,339,855	294,465
1991	1,175,428	293,107
1992	1,020,424	293,521
1993	845,626	274,286

NOTES
New England data: Landings are in metric tons of live weight,
except for univalve and bivalve molluscs, for which weights are
exclusive of their shells. Source: USDOC, various years.
Atlantic Canada data: Landings are in metric tons of live weight (fish
and shellfish only). Sources: DFO, various years, for 1977–87;
preliminary data from DFO for 1988–93.

TABLE A.2
Landings by species, Atlantic Canada and New England: 1977–93

	Atlantic Canada				New England			
	Cod	Haddock	Flounder[a]	Scallops	Cod	Haddock	Flounder[b]	Scallops
1977	237,622	26,832	111,081	116,849	34,261	12,896	44,257	7,593
1978	296,859	43,034	109,176	109,404	39,343	17,911	45,005	7,548
1979	377,986	34,599	110,186	89,488	45,065	18,997	50,036	7,277
1980	422,092	54,262	104,887	70,472	53,635	25,033	55,254	7,575
1981	439,649	57,024	105,388	89,896	45,569	25,095	52,248	8,961
1982	517,352	46,403	93,665	65,100	47,372	20,337	58,278	7,121
1983	509,052	39,777	76,963	51,289	51,017	14,770	68,898	6,115
1984	474,925	32,654	79,990	36,474	43,896	11,792	51,926	4,969
1985	480,471	37,095	89,113	47,208	37,568	6,539	41,893	4,596
1986	474,478	44,729	89,282	56,992	27,730	4,975	34,151	5,200
1987	458,204	27,511	90,066	73,855	26,807	3,027	28,757	8,155
1988	467,951	30,954	73,918	77,625	34,506	2,916	24,868	8,270
1989	422,471	26,022	70,218	92,030	35,572	1,727	21,389	9,135
1990	395,329	22,147	63,657	83,283	43,491	2,468	27,421	11,052
1991	308,325	21,949	64,266	79,373	42,019	1,838	25,774	10,708
1992	187,526	21,975	48,685	92,078	27,798	2,318	23,753	NA
1993	71,889	13,131	38,597	88,586	22,908	879	19,640	NA

NOTES

NA indicates data not available

[a]Small flatfish.

[b]Flounder landings are calculated as the sum of blackback (winter), yellowtail, and 'other flounder.'

Atlantic Canada data: Landings are in metric tons of live weight (fish and shellfish only). Sources: DFO, various years, for 1977–87; preliminary data from DFO for 1988–93.

New England data: Landings are in metric tons of live weight, except for univalve and bivalve molluscs, for which weights are exclusive of their shells. Sources: USDOC, various years; Georgianna, Dirlam, and Townsend 1993.

TABLE A.3
Atlantic Canada unit values by species: 1977–93

	Cod		Scallops		All species	
	Nominal	Real	Nominal	Real	Nominal	Real
	C$	C$	C$	C$	C$	C$
1977	0.12	0.23	0.17	0.33	0.13	0.25
1978	0.13	0.24	0.26	0.47	0.16	0.29
1979	0.15	0.24	0.38	0.62	0.19	0.31
1980	0.15	0.23	0.44	0.66	0.20	0.30
1981	0.17	0.22	0.50	0.67	0.22	0.29
1982	0.17	0.20	0.42	0.50	0.22	0.27
1983	0.17	0.19	0.63	0.71	0.25	0.29
1984	0.16	0.18	0.70	0.76	0.26	0.28
1985	0.18	0.18	0.60	0.62	0.26	0.27
1986	0.21	0.21	0.59	0.59	0.32	0.32
1987	0.32	0.31	0.58	0.56	0.40	0.39
1988	0.23	0.21	0.50	0.46	0.33	0.31
1989	0.23	0.21	0.46	0.40	0.33	0.29
1990	0.28	0.23	0.47	0.40	0.32	0.27
1991	0.33	0.26	0.46	0.37	0.39	0.31
1992	0.37	0.29	0.50	0.39	0.44	0.34
1993	0.38	0.29	0.59	0.45	0.48	0.37

NOTES
Prices in Canadian dollars per pound of live weight calculated from landings and values taken from DFO, various years, for 1977–87; preliminary data from DFO for 1988–93. Real values calculated using Canadian consumer price index, 1986 = 100.

TABLE A.4
New England unit values by species: 1977–93

	Cod		Scallops[a]		All species	
	Nominal	Real	Nominal	Real	Nominal	Real
	US$	US$	US$	US$	US$	US$
1977	0.23	0.37	1.62	2.68	0.35	0.58
1978	0.25	0.38	2.46	3.78	0.39	0.60
1979	0.29	0.40	3.28	4.52	0.43	0.59
1980	0.27	0.33	3.84	4.66	0.42	0.50
1981	0.33	0.36	3.67	4.04	0.51	0.56
1982	0.36	0.37	3.66	3.80	0.54	0.56
1983	0.34	0.34	5.45	5.47	0.61	0.61
1984	0.37	0.36	5.30	5.10	0.62	0.60
1985	0.42	0.39	4.71	4.38	0.71	0.66
1986	0.59	0.54	4.87	4.45	0.81	0.74
1987	0.75	0.66	4.13	3.63	0.94	0.83
1988	0.56	0.48	4.20	3.55	0.87	0.73
1989	0.61	0.49	3.93	3.17	0.90	0.73
1990	0.64	0.49	3.85	2.95	0.84	0.64
1991	0.80	0.59	4.04	2.97	0.92	0.68
1992	0.85	0.60	4.84	3.45	0.93	0.67
1993	0.89	0.62	5.83	4.03	0.91	0.63

NOTES
Prices calculated from landings and values taken from USDOC, various years, in U.S. dollars per pound of live weight, except scallops, for which weights are exclusive of their shells. Real values calculated using U.S. consumer price index, 1982–4 = 100.
[a]Northeast sea scallops are harvested as far south as North Carolina.

TABLE A.5
Fishery products processed in Atlantic Canada, by product form, by species: 1977–87

	Cod, haddock, flounder, and sole								
	Round or dressed		Fillets						
	Fresh total	Frozen total	Fresh total	Frozen total	Frozen blocks	Salted cod	Percent fresh	Percent frozen	Percent salt
1977	8,181	587	4,774	41,359	37,146	17,069	12	72	16
1978	9,647	1,208	7,064	50,494	42,630	19,535	13	72	15
1979	7,927	3,606	9,661	55,117	56,328	23,287	11	74	15
1980	13,696	4,848[a]	6,376	57,596	58,738	42,242	11	66	23
1981	17,446	8,052	6,134	67,182	52,811	41,046	12	66	21
1982	18,696	6,644	8,119	74,113[a]	55,586	51,402	12	64	24
1983	18,095	2,363	7,881	70,507	65,853[b]	32,364	13	70	16
1984	30,114	6,457	8,542	71,595	49,495	34,625	19	64	17
1985	44,285	9,302	8,848	70,555	47,022	34,264	25	59	16
1986	35,194	2,307	13,963	76,640	56,413	36,187	22	61	16
1987	27,204	1,326	9,678	72,847	53,545	35,501	18	64	18

NOTES
Quantities in metric tons. For all product categories other than salted cod, entries represent the sum of cod, haddock, flounder, and sole, except as noted above.
[a]Quantities of haddock confidential. These entries represent sum of cod, flounder, and sole only.
[b]Quantities of flounder and sole confidential. This entry represents sum of cod and haddock only.
Source: DFO, various years.

TABLE A.6

Number of employees in fish harvesting in Massachusetts, Nova
Scotia, and Newfoundland: 1977–93

	Massachusetts	Nova Scotia	Newfoundland
1977	NA	NA	20,243
1978	2,753	10,311	26,447
1979	3,100	10,799	32,352
1980	3,557	11,432	35,080
1981	3,346	11,388	28,587
1982	3,265	10,965	27,379
1983	3,288	12,543	28,074
1984	3,056	13,235	27,617
1985	2,833	13,958	26,564
1986	2,519	14,859	27,075
1987	2,325	15,921	29,022
1988	2,566	16,321	29,830
1989	2,188	16,012	29,176
1990	1,961	15,531	27,905
1991	1,923	16,282	24,393
1992	1,484	NA	NA
1993	1,336	NA	NA

NOTES

NA indicates data not available.

Massachusetts data: Source: USDOL, various years, for SIC 0912
(finfish).

Nova Scotia and Newfoundland data: Sources: DFO, various years,
for 1977–87; unpublished data from DFO for 1988–91.

TABLE A.7

Number of employees in fish processing plants in Massachusetts,
Nova Scotia, and Newfoundland: 1977–92

	Massachusetts	Nova Scotia	Newfoundland
1977	5,302	4,873	7,059
1978	4,999	5,551	8,161
1979	5,455	6,126	9,807
1980	5,747	7,973	13,117
1981	5,255	6,487	9,415
1982	5,182	6,693	8,470
1983	5,002	6,208	8,199
1984	4,732	5,793	8,637
1985	4,416	6,412	9,134
1986	4,018	6,778	10,285
1987	3,707	7,226	11,206
1988	3,605	7,364	11,024
1989	3,368	NA	NA
1990	3,182	NA	NA
1991	3,130	NA	NA
1992	2,653	NA	NA

NOTES

NA indicates information not available.

Massachusetts data: Source: USDOC, various years.

Nova Scotia and Newfoundland data: Sources: DFO, various years,
for 1977–87; unpublished data from DFO for 1988.

TABLE A.8
Landed values, all species, in $000s, Atlantic Canada and New England: 1977–93

	Atlantic Canada		New England	
	Real	Nominal	Real	Nominal
	C$	C$	US$	US$
1977	561,895	288,252	334,630	202,786
1978	744,184	415,999	393,420	256,510
1979	832,457	507,799	416,029	302,037
1980	759,040	510,075	397,208	327,299
1981	752,385	568,051	391,474	355,850
1982	703,912	589,174	387,480	373,918
1983	703,033	622,184	436,874	435,127
1984	648,378	599,101	417,250	433,523
1985	716,795	688,123	389,792	419,416
1986	880,145	880,145	409,299	448,592
1987	1,082,311	1,129,933	451,108	512,459
1988	935,176	1,015,601	417,284	493,647
1989	811,632	925,261	410,427	508,929
1990	783,280	936,020	415,161	542,616
1991	798,559	1,007,781	436,335	594,288
1992	768,258	984,139	430,354	603,786
1993	691,905	902,244	382,201	552,280

NOTES
Atlantic Canada data: Real values calculated using Canadian consumer price index, 1986 = 100. Sources: DFO, various years, 1977–87; preliminary data from DFO for 1988–93.
New England data: Real values calculated using U.S. consumer price index, 1982–4 = 100. Source: USDOC, various years.

TABLE A.9
Average annual salary of employees in fish harvesting in Massachusetts, Nova Scotia, and Newfoundland: 1979–93

	Massachusetts		Nova Scotia		Newfoundland	
	Nominal	Real	Nominal	Real	Nominal	Real
	US$	US$	C$	C$	C$	C$
1979	18,604	25,625	NA	NA	NA	NA
1980	15,485	18,793	NA	NA	NA	NA
1981	17,994	19,795	11,100	14,702	2,800	3,709
1982	17,079	17,698	NA	NA	NA	NA
1983	18,933	19,009	NA	NA	NA	NA
1984	17,930	17,257	NA	NA	NA	NA
1985	16,847	15,657	16,125	16,797	4,393	4,576
1986	22,179	20,236	21,227	21,227	6,181	6,181
1987	28,286	24,900	24,138	23,121	8,946	8,569
1988	24,418	20,641	18,288	16,840	7,368	6,785
1989	23,483	18,938	16,734	14,679	6,021	5,282
1990	28,624	21,901	14,900	12,469	4,300	3,598
1991	29,685	21,795	NA	NA	NA	NA
1992	30,921	22,039	NA	NA	NA	NA
1993	27,583	19,089	NA	NA	NA	NA

NOTES
NA indicates data not available.
Massachusetts data: Real values computed using U.S. consumer price index, 1982–4 = 100. Source: USDOL, various years, for SIC 0912 (finfish).
Nova Scotia and Newfoundland data: Salary represents fishing income only. Real values computed using Canadian consumer price index, 1986 = 100. Sources: Unpublished data from Nova Scotia Department of Fisheries for 1985–9; DFO 1993c for 1981 and 1990.

TABLE A.10 Real landed values by species, in $000s, Atlantic Canada and New England: 1977–93

	Atlantic Canada					New England				
	Cod	Haddock	Flounder[a]	Total cod, haddock, and flounder	Scallops	Cod	Haddock	Flounder[b]	Total cod, haddock, and flounder	Scallops
	C$	C$	C$	C$	C$	US$	US$	US$	US$	US$
1977	120,357	22,238	45,684	188,279	85,949	28,210	15,297	65,383	108,889	44,750
1978	154,530	33,295	43,501	231,326	113,564	33,000	19,431	72,988	125,419	62,783
1979	198,802	26,218	45,600	270,620	122,051	39,438	24,387	68,778	132,603	72,481
1980	212,613	39,396	44,034	296,043	101,929	38,693	26,000	59,517	124,210	77,825
1981	215,746	33,034	38,532	287,313	131,926	36,393	24,218	57,531	118,142	79,759
1982	230,890	27,572	32,774	291,237	72,326	38,741	23,123	66,079	127,943	59,546
1983	211,809	27,452	26,362	265,623	79,999	38,080	19,045	75,377	132,502	73,772
1984	185,366	24,503	27,666	237,535	61,023	34,786	17,663	75,482	127,932	55,877
1985	193,509	28,990	31,315	253,814	64,761	32,658	12,588	69,544	114,790	44,351
1986	214,880	37,159	37,725	289,764	74,303	32,976	9,955	63,572	106,504	50,935
1987	310,484	32,380	44,261	387,125	90,805	38,890	7,502	65,238	111,629	65,364
1988	220,367	26,048	33,980	280,395	78,745	36,298	5,943	52,191	94,432	64,729
1989	190,962	22,227	34,129	247,318	81,516	38,526	3,660	45,348	87,534	63,828
1990	203,406	20,254	30,321	253,981	72,745	46,923	4,565	44,850	96,339	71,771
1991	179,537	24,324	27,881	231,742	64,309	54,400	3,363	42,510	100,274	70,024
1992	119,431	23,610	22,942	165,984	78,559	37,073	3,979	39,752	80,803	NA
1993	45,963	16,102	18,143	80,209	88,741	31,111	1,849	37,464	70,424	NA

NOTES: NA indicates data not available.

[a] Small flatfish

[b] Flounder landings are calculated as the sum of blackback (winter), yellowtail, and 'other flounder' landings.

Atlantic Canada data: Landed values are in C$000s. Real values calculated using Canadian consumer price index, 1986 = 100. Sources: DFO, various years, 1977–87; preliminary data from DFO for 1988–93.

New England data: Landed values are in US$000s. Real values calculated using U.S. consumer price index, 1982–84 = 100. Sources: USDOC, various years; Georgianna, Dirlam, and Townsend (1993)

TABLE A.11
Nominal landed values by species, in C$000s, Atlantic Canada: 1977–93

	Cod	Haddock	Flounder[a]	Total cod, haddock, and flounder	Scallops
	C$	C$	C$	C$	C$
1977	61,743	11,408	23,436	96,587	44,092
1978	86,382	18,612	24,317	129,311	63,482
1979	121,269	15,993	27,816	165,078	74,451
1980	142,876	26,474	29,591	198,941	68,496
1981	162,888	24,941	29,092	216,921	99,604
1982	193,255	23,078	27,432	243,765	60,537
1983	187,451	24,295	23,330	235,076	70,799
1984	171,278	22,641	25,563	219,482	56,385
1985	185,769	27,830	30,062	243,661	62,171
1986	214,880	37,159	37,725	289,764	74,303
1987	324,145	33,805	46,208	404,158	94,800
1988	239,319	28,288	36,902	304,509	85,517
1989	217,697	25,339	38,907	281,943	92,928
1990	243,070	24,204	36,233	303,507	86,930
1991	226,576	30,697	35,186	292,459	81,158
1992	152,991	30,245	29,389	212,625	100,634
1993	59,936	20,997	23,659	104,592	115,718

[a]Small flatfish.
Sources: DFO, various years, for 1977–87; preliminary data from DFO for 1988–93.

TABLE A.12
Nominal landed values by species, in US$000s, New England: 1977–93

	Cod	Haddock	Flounder[a]	Total cod, haddock, and flounder	Sea Scallops[b]
	US$	US$	US$	US$	US$
1977	17,095	9,270	39,622	65,987	27,119
1978	21,516	12,669	47,588	81,773	40,934
1979	28,632	17,705	49,933	96,270	52,621
1980	31,883	21,424	49,042	102,349	64,128
1981	33,081	22,014	52,296	107,391	72,501
1982	37,385	22,314	63,766	123,465	57,462
1983	37,928	18,969	75,075	131,972	73,477
1984	36,143	18,352	78,426	132,921	58,056
1985	35,140	13,545	74,829	123,514	47,722
1986	36,142	10,911	69,675	116,728	55,825
1987	44,179	8,522	74,110	126,811	74,253
1988	42,941	7,030	61,742	111,713	76,574
1989	47,772	4,538	56,232	108,542	79,146
1990	61,329	5,967	58,619	125,915	93,805
1991	74,093	4,581	57,899	136,573	95,372
1992	52,013	5,582	55,772	113,367	NA
1993	44,956	2,672	54,135	101,763	NA

NOTES
NA indicates data not available.
[a]Flounder landings are calculated as the sum of blackback (winter), yellowtail, and 'other flounder' landings.
[b]Landed values of sea scallops calculated as the product of unit values for all northeast sea scallops and landed quantities for New England.

TABLE A.13

Cod products processed in Nova Scotia, percentage by processing method: 1977–87

	Frozen quantity	Percent of total	Fresh quantity	Percent of total	Salted quantity	Percent of total	Total quantity
1977	12,105	58.2	3,088	14.8	5,605	26.9	20,798
1978	15,243	58.2	3,743	14.3	7,220	27.6	26,206
1979	21,106	57.5	4,744	12.9	10,827	29.5	36,677
1980	19,939[a]	46.2	6,851	15.9	16,347	37.9	43,137
1981	19,024	39.4	7,134	14.8	22,146	45.8	48,304
1982	28,393	44.3	5,804	9.1	29,932	46.7	64,129
1983	30,570	55.7	11,356	20.7	12,933	23.6	54,859
1984	31,747	49.4	14,929	23.2	17,564	27.3	64,240
1985	30,872	42.3	22,301	30.6	19,816	27.1	72,989
1986	23,911	34.9	24,738	36.1	19,912	29.0	68,561
1987	25,297	38.4	21,046	31.9	19,603	29.7	65,946

NOTES

Quantity in metric tons of product weight.

[a]Quantities of frozen round and dressed cod are confidential for this year. This entry represents the sum of frozen fillets and frozen blocks only. All other entries represent the sum of frozen round and dressed cod, frozen fillets, and frozen blocks.

Source: DFO, various years.

TABLE A.14
Cod products processed in Newfoundland, percentage by processing method: 1977–87

	Frozen quantity	Percent of total	Fresh quantity	Percent of total	Salted quantity	Percent of total	Total quantity
1977	29,089	69.3	4,051	9.7	8,831	21.0	41,971
1978	36,546	72.3	4,102	8.1	9,920	19.6	50,568
1979	51,427	78.3	4,340	6.6	9,937	15.1	65,704
1980	54,392	69.7	6,081	7.8	17,595	22.5	78,068
1981	61,868	78.5	4,034	5.1	12,926	16.4	78,828
1982	74,528	72.7	12,608	12.3	15,319	15.0	102,455
1983	75,905	77.1	9,584	9.7	12,998	13.2	98,487
1984	67,085	74.8	11,796	13.1	10,843	12.1	89,724
1985	60,122	76.9	7,946	10.2	10,128	13.0	78,196
1986	72,290	79.1	6,986	7.6	12,106	13.2	91,382
1987	66,595	76.8	7,764	9.0	12,361	14.3	86,720

NOTE: Quantity in metric tons of product weight.
Source: DFO, various years.

TABLE A.15
Canadian groundfish exports to the United States: 1977–93

	Fresh whole cod	Percent of total	Fresh whole haddock	Percent of total	Fresh flounder fillets	Percent of total	Fresh cod fillets	Percent of total	Fresh haddock fillets	Percent of total	Frozen[a] fillets	Percent of total	Frozen[a] blocks	Percent of total	Total
1977	1,290	1.9	2,397	3.5	622	0.9	1,591	2.3	634	0.9	27,547	40.2	34,430	50.3	68,511
1978	801	1.0	2,454	3.2	355	0.5	1,520	2.0	883	1.1	34,843	45.0	36,533	47.2	77,389
1979	1,383	1.5	2,640	2.8	207	0.2	2,812	3.0	1,072	1.1	38,791	41.2	47,229	50.2	94,134
1980	1,896	2.0	3,588	3.7	314	0.3	2,278	2.3	992	1.0	38,919	40.1	48,995	50.5	96,982
1981	2,766	2.6	6,698	6.2	513	0.5	2,392	2.2	1,579	1.5	51,607	48.0	42,007	39.1	107,562
1982	3,433	2.9	9,295	7.9	434	0.4	3,123	2.7	1,186	1.0	57,144	48.6	42,935	36.5	117,550
1983	5,382	4.3	11,249	9.0	414	0.3	4,276	3.4	1,303	1.0	53,457	42.7	49,069	39.2	125,150
1984	11,891	9.0	14,447	10.9	909	0.7	6,159	4.7	904	0.7	56,987	43.1	40,934	31.0	132,231
1985	15,002	10.8	17,897	12.8	2,363	1.7	12,268	8.8	1,210	0.9	47,005	33.7	43,599	31.3	139,344
1986	13,263	8.8	18,622	12.3	2,865	1.9	15,670	10.4	1,725	1.1	46,794	31.1	51,827	34.3	150,946
1987	NA		NA		NA		NA		NA		NA		NA		NA
1988	11,053	8.7	8,028	6.3	3,080	2.4	15,406	12.1	1,381	1.1	31,956	25.1	56,368	44.3	127,272
1989	10,000	8.1	8,018	6.5	1,905	1.5	9,759	7.9	1,004	0.8	44,067	35.7	48,529	39.4	123,282
1990	5,239	4.9	8,554	7.9	1,794	1.7	5,671	5.3	678	0.6	36,852	34.1	49,147	45.5	107,935
1991	3,786	4.4	8,288	9.5	1,888	2.2	3,632	4.2	608	0.7	31,592	36.4	37,104	42.7	86,898
1992	5,365	9.4	9,236	16.2	1,662	2.9	2,772	4.9	776	1.4	20,684	36.3	16,429	28.9	56,924
1993	5,052	13.7	6,283	17.0	1,490	4.0	2,167	5.9	681	1.8	13,917	37.7	7,297	19.8	36,887

NOTES

Quantities in metric tons. NA indicates information not available.

[a] Calculated as sum of cod, haddock, flounder and sole frozen fillets/blocks.

Sources: Statistics Canada, various years, for 1975–86: Statistics Canada, International Trade Division, unpublished data, 1988–90; U.S. Bureau of the Census, 1991–2; U.S. Bureau of the Census, 1993.

TABLE A.16
Average real unit value (average price) of Canadian cod exports
to the United States: 1977–93

	Fresh whole cod	Fresh fillets	Frozen fillets	Frozen blocks
	C$	C$	C$	C$
1977	0.86	2.16	1.98	1.68
1978	0.84	2.19	2.01	1.76
1979	0.75	2.08	1.95	1.78
1980	0.76	1.96	1.86	1.64
1981	0.72	1.91	1.86	1.49
1982	0.66	1.77	1.80	1.37
1983	0.64	1.65	1.76	1.41
1984	0.55	1.67	1.72	1.21
1985	0.63	1.73	1.78	1.30
1986	0.71	2.11	2.07	1.61
1987	NA	NA	NA	NA
1988	0.94	1.92	1.99	1.59
1989	0.71	1.74	1.82	1.36
1990	0.60	1.80	1.94	1.64
1991	0.63	1.99	2.13	1.71
1992	0.73	2.10	2.24	1.69
1993	0.84	2.26	2.30	1.41

NOTES
Unit values expressed in real Canadian dollars per pound. Real
values computed using Canadian consumer price index,
1986 = 100. Export values for 1991–3 converted from U.S. to
Canadian dollars using annual average exchange rates published
in International Monetary Fund, *International Financial Statistics*.
February 1994. NA indicates information not available.
Sources: Statistics Canada, various years, for 1975–86; Statistics
Canada, International Trade Division, unpublished data, for
1988–90; U.S. Bureau of the Census, 1991–2; U.S. Bureau of the
Census 1993.

Notes

CHAPTER 2

1 The New England fleet consists of those vessels based in the states of Maine, New Hampshire, Massachusetts, Rhode Island, and Connecticut. The Atlantic Canada fleet covers vessels based in the provinces of New Brunswick, Quebec, Prince Edward Island, Nova Scotia, and Newfoundland.
2 Approximately 20 percent of the fishermen employed in Gloucester in 1886 were Canadian migrants (Innis 1940). For New England as a whole, an estimated 2,250 of the 14,000 fishermen were Canadian (Innis 1940).
3 During the early to mid-1980s part-time harvesting employment in Nova Scotia rose from 39 percent to 47 percent, while it fell slightly in Newfoundland from 53 percent to 47 percent (DFO, *Annual Statistical Review,* various years).

CHAPTER 3

1 The increase in small vessels in the 1990–92 period is a statistical artifact caused by improved survey techniques. These data are collected by NMFS through port agents who count active vessels. There has been no required vessel reporting until the recently enacted Amendment #5. In the 1990–92 period, four additional ports in Maine were covered, thus increasing the reported number of smaller vessels, even though the active number of smaller vessels fishing did not increase, but on the contrary, decreased during this period.
2 This pattern reflects the decline in the number of smaller offshore vessels (50–150 tons) to below their 1980 level and the steady rise in the number of large vessels during this period. The average large vessel produces a little over two and one-half times the revenue of the smaller offshore vessels

($575,000 compared with $227,000 in 1989; USDOC 1991b).

3 Data on the number of New England vessels between 5 and 10 tons are not available, nor are counterpart data for Canada. Consequently, a direct comparison between exactly the same size smaller vessels in each country is not possible. A comparison of tables 3.1 and 3.2, however, shows that the number of vessels 50 tons and greater in the Canadian fleet is only a little over one-third larger than the number in the New England fleet.

4 If applied to the New England fleet, this definition would cover all of the smaller offshore vessels (50–150 tons), as well as the inshore fleet.

5 In Nova Scotia, the nearshore fleet has been further refined into 35–45 foot and 45–65 foot categories, as discussed below, in chapter 4.

6 It is stated elsewhere (Weeks and Mazany 1983) that by far the majority of inshore fish are frozen into blocks. This may be the case in Newfoundland, but more recent interviews with government officials, processors, and fishermen in Nova Scotia produced consistent statements that inshore fish from this area are destined largely for the fresh-fish markets.

7 Within New England, Massachusetts accounts for around 93 percent of all groundfish processing, the remainder being processed in Rhode Island and Maine.

8 The salt-fish market was also quite concentrated, and one firm, the Canadian Saltfish Corporation, accounted for more than 30 percent of the sales. This firm is required by law to purchase all salt-fish produced by fishermen in Newfoundland, Labrador, and Quebec. In 1981 the four largest salt-fish companies accounted for 73 percent of all Atlantic Canadian salt-fish exports.

9 Most of these equity purchases were used to retire debts owed to a Canadian bank. Management of the two companies remained in private hands, however, and the federal government was committed to selling its shares as soon as they were profitable. Although Fisheries Product International was fully privatized in 1987 (FPI 1991), the federal and provincial governments retain their non-voting shares in National Sea Products.

10 This restructured, large-scale sector was also supposed to ensure a more stable demand for fish from small processors and wholesalers, the vast majority of whom purchase their fish from the inshore fleet and are too small to afford frozen-fish processing equipment. These small processors and wholesalers would gain the option to freeze by marketing their whole fish through the large processors whenever doing so was more profitable than selling them fresh.

CHAPTER 4

1 There is a large literature that criticizes traditional economic efficiency analy-

sis for its inability to deal with these distributional conflicts (Bromley and Bishop 1977; Wilson 1982; Bannister 1989).

2 For a recent review of this literature, see Anderson (1986), Townsend (1990), and Crutchfield (1979).

3 In precise terms, as the costs of fishing are increased by management, the MEY level of effort is reduced, because although the costs of fishing are increasing, the benefits as measured by the revenue from the catch are not. Therefore, less effort is required to reach the point where the additional value of the harvest is equivalent to the additional cost of harvesting. This conclusion, of course, is based on the presumption that managers can determine the efficient level of harvesting effort. Some authors have questioned the ability to predict changes in fish stocks in response to restrictions or to fine-tune controls on fishing effort. They have suggested that open-access restrictions are a realistic compromise in the face of alternative, more efficient effort restrictions that must rely for their success on more accurate knowledge of stock behavior than currently exists (Wilson et al. 1991; Wilson 1992).

4 The chief implementation problems that have been raised concerning transferable quotas are the difficulties of enforcement and of resolving what should be done with the incidental harvesting (or 'by-catch') of species where the quota has already been filled.

5 There have been attempts (most recently in 1980) by some provinces to acquire some independent management authority, but this decentralization of regulatory policy has been resisted. There is no agreement among the provinces about how a decentralized system would operate, and the large fish-processing companies have lobbied against being subject to multiple levels of administrative authority (Copes 1982; Bannister 1989).

6 Accompanying these goals were recommendations for aggressive economic development programs to be coordinated with the relevant provinces; many of these recommendations were eventually adopted (Bannister 1989).

7 The precise definition is 'the level of fishing mortality at which the marginal increase in yield by adding one more unit of fishing effort is 10 percent of the increase in yield by adding the same unit of effort in a lightly exploited stock. It is an internationally accepted level of fishing dictated by practical experience in stock management' (Atlantic Fisheries Service, 1986, 4). In the case of cod, haddock, and pollock, annual harvests must be limited to 'about 20 percent of individuals in age groups which are fully recruited to the fishery' (Halliday, Peacock, and Burke 1992, 413).

8 If the TAC changed by more than 15 percent from the previous year, these allocations could be altered (Atlantic Fisheries Service 1986).

9 The FRCC issued its first recommendations for the 1993 fishing season in

August of that year (FRCC 1993a), and the 1994 recommendations were issued in November 1993 (FRCC 1993b).

10 One other proposal for change being seriously considered in 1994 is the creation of an Atlantic Coast Fisheries Board, to be charged with determining how the TACs set by the minister would be allocated among the various industry constituents and who would get fisheries licenses (DFO 1993a). The allocation process and criteria would thus be opened to the public; currently allocation decisions are made within DFO.

11 Moreover, from the standpoint of economic efficiency, technological choice that reduced the cost of catching fish should have been allowed.

12 Ironically, much of this expansion of the nearshore fleet was either government financed through traditional vessel subsidy programs or was made possible by the success of ICNAF during the early 1970s in rebuilding the haddock stocks, which generated high incomes for inshore fishermen during the late 1970s and early 1980s (Bannister 1989; Halliday, Peacock, and Burke 1992).

13 Horsepower and the use of modern fishing technologies, however, are still not restricted.

14 The expansion of the nearshore fleet has been interpreted by some as threatening to substitute specialized vessels incapable of quickly converting to alternative fisheries when stocks are low (the nearshore fleet) for traditional inshore vessels that are quite flexible and therefore readily adapt to changing market and biological conditions (Bannister 1989; Apostle and Barrett 1992a).

15 A fleet of eight nearshore vessels owned by a southwest Nova Scotia fishing family, for example, was reported to be capable of catching 25 percent of the 1989 inshore quota for all of southwest Nova Scotia (Cameron 1990).

16 The system is also used by Canadian managers for other Atlantic species, including herring, shrimp, lobster, tuna, and, on an informal basis, scallops.

17 A recent government task force report concluded: (1) there was substantial overcapacity in the inshore fishery, largely owing to the inefficiencies of the limited entry licensing regulations; (2) the substantial overcapacity in the inshore sector needed to be addressed immediately if the fish stocks were to be rejuvenated; and (3) the inshore overcapacity problem should be remedied not by the transfer of quotas from the offshore to the inshore sector, but instead by restructuring the inshore fleet (Hache 1989).

18 Although this regulatory option still gives an incentive for individual vessels to catch fish as rapidly as possible in order to capture as large a share of the quota as possible, this was not felt to be a problem in the immediate future, since there is not substantial overcapacity in this sector (Hache, 1993). Recently, however, these vessels have been allowed to continue fishing after the quota was reached (Burke et al. 1994).

19 This fleet will be allowed to contract with the large processors, which would be attracted by the smaller vessels' greater harvesting efficiency compared with their own larger, older vessels, to catch some of the offshore enterprise quota.

20 NMFS is also responsible for the development of regulations to carry out the intent of the plan, overseeing the enforcement of the regulations by the Coast Guard and providing administrative support for the licensing of vessels and collecting catch statistics.

21 For example, a recent task force studying the New England groundfish industry concluded that the 1990 groundfish management plan was essentially the same as the Interim Plan implemented in 1982 (Massachusetts Task Force 1990, 19)!

22 The report was issued by the Stock Assessment Review Committee, which is responsible for annually assessing the New England region groundfish stocks.

23 The one-day layover for every two days of fishing applies each year in addition to the declared days out of the fishery. Each option requires the vessels not to fish for one twenty-day block of time during spawning season (1 March–31 May). The reduced fishing time limits will be enforced by electronic vessel tracking systems for vessels choosing the 10 percent reduction option, and all vessels must fax or phone NMFS at the start and end of each trip.

24 If this effort reduction is not successful in increasing stocks within three years, the plan enables the council to impose more stringent effort restrictions. NMFS has already expressed concern that the exemption for vessels 45 feet or less will undermine the gains from the restrictions on the main fleet.

25 This bill also includes the imposition of a landing fee on all fish in order to support improved enforcement, data collection, scientific research, and conservation and management.

26 When the money is used, the depreciable value of the new or reconstructed vessel is reduced by the amount of the CCF fund contribution. Thus, the loan is repaid in the form of lower depreciation deductions allowed against future income for tax payment purposes.

27 The 1994 number was provided by Elizabeth Brown, CCF Department, National Marine Fisheries Service. The only other nonemergency direct U.S. federal assistance program to the fishing industry compensates fishermen for gear damage from other vessels and is financed from fines assessed on owners of foreign fishing vessels seized in U.S. waters. A little over 200 claims totalling slightly more than $2 million were submitted by New England fishermen between 1979 and 1983 (USITC 1984).

28 An additional $4 million has been appropriated for 1995.

29 Note that if costs were zero, MSY would have to be the level of effort where total revenue minus total costs is the largest (MEY), since total revenue is at a maximum at MSY. Another way of understanding this point is that the marginal revenue of additional effort, which is zero, would be equal to the marginal cost of additional effort. As long as costs are not zero, marginal revenue will always be less than marginal costs at this level of effort, thus indicating overfishing.

30 Since increasing prices or increases in productivity cause the revenue curves to shift outward and falling costs shift the cost curves down, MEY and MSY would occur at higher levels of effort also.

31 OA can fall below MSY, since the average costs can equal average revenues below the level of effort that generates maximum revenue, but it can never fall to MEY.

32 The marginal cost curve is rising because of the regulations and thus must intersect the same marginal revenue curve at a lower level of fishing effort.

CHAPTER 5

1 Under the lay system, the income of a crew member depends primarily on (1) the volume and selling price of the catch, (2) the operating costs of the vessel, and (3) the number of crew among whom the boat's income must be shared. It contains elements of a piece-rate system in that income is tied, in part, to the volume of catch, so that there are incentives for effort and rewards for productivity. Since the lay system involves income sharing, there is an obvious tradeoff between the size of the crew and the income of individual crew members.

2 They include the United Fishermen and Allied Workers Union, the Newfoundland Fishermen, Food and Allied Workers Union (chartered by the Canadian Automobile Workers), the Maritime Fishermen's Union, the Canadian Food and Allied Workers' Union (an affiliate of the Amalgamated Meatcutters and Butcher Workmen's union), and the Canadian Brotherhood of Railway, Transport, and General Workers Union. Union activity on behalf of the small inshore fleet is directed more at political activity, although the unions sometimes become involved in the negotiation of minimum fish prices and preferential purchasing agreements with processors (Macdonald 1980). The Maritime Fishermen's Union provides political representation for smaller fishermen in northern Nova Scotia, whereas many of the southwest Nova Scotia independents are nonunion and tend to join industry associations (Clement 1984; Steinberg 1974). The nearshore fleet is the least organized.

3 Paradoxically, unionization in New Bedford remains far more common on kinship vessels than it is on other vessels, mainly because of the attraction of union-sponsored health and pension plans, rather than the economic benefits of bargaining power. While contract provisions are strictly enforced on non-kinship vessels, there is more informality and flexibility in work rules on the kinship vessels, since the rights and responsibilities of 'family' have become far more important union safeguards.

4 For example, collective bargaining in Canada has been used to clarify the definition of job duties for specialized classifications (engineer, cook, mate, deckhand) and has altered the handling of shift assignments and 'call-in' pay (Clement 1984; Binkley 1990). Similarly, collective bargaining in New England after the Second World War led to the detailed regulation of hiring, staffing, and work, and it provided seniority and job security protections for crew members who would otherwise be subject to the unilateral authority of captains and owners (White 1954).

5 What little remains of fresh-fish processing in Gloucester is nonunion, and barely any union organization remains in New Bedford since a failed strike in 1981. The principal Gloucester unions in frozen processing are the United Food and Commercial Workers International Union and the Amalgamated Meatcutters and Butcher Workmen of North America. In New Bedford, the Seafood Workers Union (affiliated with the International Longshoremen's Association) is the principal union.

6 The Canadian experience suggests that these improvements in working conditions, along with increased compensation, have helped to reduce the high rate of labor turnover that accompanied the introduction of corporate capitalism into plants previously operating under the more relaxed arrangements of paternalistic capitalism (Apostle et al. 1985, 1992).

7 Although there are few data on labor mobility among vessels, one study of unemployed fishermen showed, in Gloucester (a kinship port), a much stronger attachment to a single vessel than is true in New Bedford (a mixed-kinship and corporate capitalism port). In Gloucester, over 60 percent of the fishermen had worked on a single boat and only 13 percent had worked for three or more employers, as would be expected in a kinship port, whereas in New Bedford only about one-third of the fishermen had worked on only one boat and almost half had worked for three or more employers during the year (Doeringer, Moss, and Terkla 1986).

8 See chapter 4, above. Understanding the overall effect of the regulatory process on employment systems, however, is complicated by the facts that fish harvested under the TAC allowance for the smaller inshore vessels may be processed in the corporate sector and that independent vessels may fish under corporate sector quotas.

9 The increase in wage earners may also be influenced by the strategic use of the regular unemployment insurance system, to which self-employed fishermen do not have access (DFO 1993c.)

10 This view is consistent with the observation that productivity in processing seems to rise during times of crisis as a result of management and union efforts (Kirby 1982).

11 These differences, however, overstate unit labor cost differences because of differences between inshore and offshore vessels in the number of days fished.

CHAPTER 6

1 There were two sources of data on groundfish imports until 1990 – United States import data and Canadian export data. After 1990 the two series are identical. Prior to 1990 both series of data were recorded in a similar manner, although the exact makeup of each category differed in the two series, as did the date of record. The two series differ significantly throughout the 1977–90 period, however, with the United States import data consistently underpredicting the Canadian export data except at low quantities.

Since imports are concentrated at the beginning of each quarter to take advantage of the lower import duties on below-quota cod and haddock fillets, a three-month moving average was created in order to smooth out monthly fluctuations and to enable a comparison between the two data series. Simple correlation coefficients between the moving averages for the two series are 0.91 for fresh cod fillets and 0.82 for fresh haddock fillets, suggesting strong similarities in the series for cod and haddock, the major species traded. Furthermore, an inspection of both series indicates that, although quantities differed significantly, meaningful trends and fluctuations are very similar in both series.

Consequently, in the discussion of trends of fresh whole and filleted fish exports prior to 1990, the Canadian export series are used because they provide the most published data. Unfortunately, 1987 data are not available. U.S. import data are used for the econometric model, however, since most similar econometric studies have used this series and these data are complete for the entire 1977–90 period.

In the text, we adopt the convention of referring to the trade flow from the Canadian perspective. Thus, fresh-fish trade is referred to as 'exports' rather than 'imports.'

2 Over 90 percent of Canadian frozen-fillet and frozen-block exports were sent to the United States in 1986 (Statistics Canada 1986). Canada is the largest

supplier of frozen blocks to the United States, and its market share has been rising in recent years. In 1990 Canadian exports to the United States represented 44 percent of the quantity and 50 percent of the value of total U.S. block imports, over twice their 1977 level and up from 37 percent in 1983 (USDOC, *Fisheries of the United States*, 1991; USITC 1984; Kirby 1982). Other major exporters include Denmark, Iceland, and Korea.

3 Canada is a significant supplier of other groundfish species, but in recent years as Canadian stocks have fallen, other countries, such as Iceland and Argentina, have also become important suppliers of species such as haddock and flounder.

4 Currency changes in other countries exporting frozen fish may also have influenced the expansion of Canadian fresh-fish exports. For example, in real terms the Icelandic *krone* fell much more than the Canadian dollar, depreciating 27 percent relative to the U.S. dollar during the 1979–83 period. Most of this depreciation took place between 1981 and 1983. Although one might expect that this change would force Canadian producers from frozen to fresh markets, Canada's share of frozen cod-fillet imports actually increased relative to Iceland's import share between 1981 and 1982 and held steady through 1983. This paradoxical result may be partially explained by the segmented nature of the frozen-fish market in which middle-quality Canadian fish are not in direct competition with most high-quality Icelandic imports. Nevertheless, the more rapid increase in Canadian fish prices due to the exchange rate did not result in a large substitution of Icelandic for Canadian frozen-fish products.

5 The USITC investigation concluded that subsidies to the fishing industry are much broader in scope in Canada, Iceland, and Norway than they are in the United States. Moreover, given the differences in industry size among the countries, the overall level of assistance is lower in the United States. The USITC describes major loan programs and some of the major harbor development programs in the United States, but it is incomplete. For example, it makes no mention of the disproportionate subsidies U.S. fishermen receive from the U.S. unemployment insurance program. In 1981 almost $4 million of unemployment insurance payments were made to Massachusetts fishermen, representing almost 6 percent of total wages and salaries covered in that sector. Fishermen received $2.44 in benefits for every payroll tax dollar contributed, compared with an average of $0.59 per dollar for all workers in the Commonwealth. For more detail, see Doeringer, Moss, and Terkla (1986a).

6 Until January 1986 the duty on fresh whole flounder was $0.005/lb; on fresh cod and haddock fillets it was $0.01875/lb below quota and $0.025/lb above

quota prior to 1980, reducing in stages to $0.01875/lb above quota by 1987. The quota per quarter for 1983 was 12.5 million pounds, less than 1 percent of total imports of each product form. All of these duties amounted to less than 1 percent of the import unit value of each product, and all were phased out in 1993 as part of the United States – Canada Free Trade Agreement (NEFMC 1993).

7 The only published study dealing with the relationship between tariffs and fisheries imports (a simulation study completed prior to the setting of the countervail duty) does not support a large effect of duties on trade flows (Crutchfield 1985). The study concludes that Canadian suppliers would absorb the brunt of a hypothetical 20 percent tariff, but that exports would not be reduced enough to limit their current downward pressure on U.S. ex-vessel prices. Since the actual tariff imposed was far smaller than the one hypothesized in this study, any impact on fresh-fish exports is likely to be minimal.

Unfortunately, the study does not consider whole fresh fish on which the countervail duty was imposed but which had only recently been significant in terms of quantity imported. In addition, other important sources of imports, such as Iceland and Norway, are not included in the model. Since these countries are major competitors with Canada in the U.S. fish import market and in worldwide frozen-fish markets, some important interactive effects may be missed by ignoring them in this analysis. Even then, the small magnitude of the tariff that was imposed makes it unlikely that it had much impact on the volume of fresh-fish imports.

8 This model differs from Hogan and Georgianna (1989) in that it examines a longer time period (1978–90) and examines fresh-fillet imports as well as whole fish. While the focus of Hogan and Georgianna is on the role of pro-cessing-capacity utilization in influencing U.S. processors' import demand, we are interested in a broader range of influences and do not attempt to measure directly the role of U.S. processing capacity. We incorporate a time trend and seasonality adjustments explicitly in our model, which, along with a variable representing U.S. groundfish landings, serve as proxies for the influence of processing-capacity utilization as well as the other institu-tional changes we are trying to measure.

9 Salt-fish prices were not included because of the lack of readily available monthly data over the time period tested and the minimal importance of salt-fish as an alternative market for fish of sufficiently high quality to be sold in fresh markets.

References

Acheson, James M. (1984) 'Government Regulation and Exploitive Capacity: The Case of the New England Ground Fishery.' *Human Organization 43*, 319–29
– (1988) *The Lobster Gangs of Maine* (Hanover, NH: University Press of New England
Akerlof, George (1984) 'Gift Exchange and Efficiency Wages: Four Views.' *American Economic Review* 74 (May) 79–83
Andersen, Raoul (1972a) 'Hunt and Deceive: Information Management in Newfoundland Deep-Sea Trawler Fishing.' In *North Atlantic Fishermen*, ed. R. Andersen and C. Wadel. St John's: Memorial University of Newfoundland
– ed. (1979a) *North Atlantic Maritime Cultures: Anthropological Essays on Changing Adaptations*. The Hague: Mouton
– (1979b) 'Public and Private Access Management in Newfoundland Fishing.' In Andersen, ed. (1979a)
Andersen, Raoul, and Geoffrey Stiles (1973) 'Resource Management and Spatial Competition in Newfoundland Fishing: An Exploratory Essay.' In *Seafarer and Community: Towards a Social Understanding of Seafaring*, ed. Peter H. Fricke. Totowa, NJ: Rowman and Littlefield
Andersen, Raoul, and Cato Wadel, eds (1972b) *Anthropological Essays on Modern Fishing*, Toronto: University of Toronto Press
Anderson, Lee G., (1986) *The Economics of Fisheries Management*, rev. and enl. ed. Baltimore: Johns Hopkins University Press
Anthony, Vaughn C. (1990) 'The New England Groundfish Fishery after 10 Years under the Magnuson Fishery Conservation and Managenent Act.' *North American Journal of Fisheries Management* 10, 175–84
Antler, Ellen, and James Faris (1979) 'Adaptation to Changes in Technology and Government Policy: A Newfoundland Example (Cat Harbour).' In Andersen, ed. (1979a)

Apostle, Richard, and Gene Barrett (1987) 'Nova Scotia Fish Processing and the New England Market.' In *Resource Economies in Emerging Free Trade, Proceedings of a Maine/Canadian Trade Conference*, ed. V. Konrad, et al. (Orono, ME: University of Maine Press

– (1992a) *Emptying Their Nets: Small Capital and Rural Industrialization in the Nova Scotia Fishing Industry*. Toronto: University of Toronto Press

– (1992b) 'Surplus Labor.' In Apostle and Barrett (1992a)

– (1992c) 'Communities and Their Social Economy.' In Apostle and Barrett (1992a)

Apostle, Richard, Gene Barrett, Anthony Davis, and Leonard Kasdan (1985) 'Land and Sea: The Structure of Fish Processing in Nova Scotia.' Working Paper No. 1-0284, Gorsebrook Research Institute for Atlantic Canada Studies, St Mary's University, Halifax, NS, May

– (1992) 'Small, Competitive, and Large: Fish Plants in the 1980s.' In Apostle and Barrett (1992a)

Apostle, Richard, Don Clairmont, and Lars Osberg (1986) 'Economic Segmentation and Politics.' *American Journal of Sociology* 91, 905–31

Apostle, Richard, and Svein Jentoft (1991) 'Nova Scotia and North Norway Fisheries: The Future of Small-Scale Processors.' *Marine Policy* (March), 100–10

Apostle, Richard, R.L. Kasdan, and A. Hanson (1985) 'Work Satisfaction and Community Attachment among Fishermen in Southwest Nova Scotia.' *Canadian Journal of Fisheries and Aquatic Sciences* 42, 256–67.

Atlantic Fisheries Service, Department of Fisheries and Oceans (1986) 'Enterprise Allocations for the Atlantic Offshore Groundfish Fisheries – 1986.' Supply and Services Canada, October

Bannister, Ralph K. (1989) 'Orthodoxy and the Theory of Fishery Management.' Master's thesis, Atlantic Canada Studies, St Mary's University, Halifax, NS, October

Baran, Paul A., and Paul M. Sweezy (1966) *Monopoly Capital: An Essay on the American Economic and Social Order*. New York and London: Monthly Review Press

Barber, Pauline (1992) 'Household and Workplace Strategies in "Northfield."' In Apostle and Barrett (1992a)

Barrett, L. Gene (1980) 'Perspectives on Dependency and Underdevelopment in the Atlantic Region.' *Canadian Review of Sociology and Anthropology* 17, 273–86

– (1992a) 'Mercantile and Industrial Development to 1945.' In Apostle and Barrett (1992a)

– (1992b), 'Post-war Development'. In Apostle and Barrett, (1992a)

Barrett, Gene, and Anthony Davis (1984) 'Floundering in Troubled Waters: The Political Economy of the Atlantic Fishery and the Task Force on Atlantic Fisheries.' *Journal of Canadian Studies* 19, 125–37

Binkley, Marian (1990) 'Work Organization among Nova Scotian Offshore Fishermen.' *Human Organization* 49, 395–405

Boeri, David, and James Gibson (1976) *Tell It Good-bye Kiddo: The Decline of the New England Offshore Fishery.* Camden, ME: International Marine Publishing Company

Boyer, Robert (1988) 'Wage/Labour Relations, Growth, and Crisis: A Hidden Dialectic.' In *The Search for Labour Market Flexibility*, ed. Robert Boyer. Oxford: Clarendon Press

Braverman, Harry (1974) *Labor and Monopoly Capital: The Degradation of Work in the Twentieth Century.* New York and London: Monthly Review Press

Bromley, Daniel W., and R.C. Bishop (1977) 'From Economic Theory to Fisheries Policy: Conceptual Problems and Management Prescriptions.' In *Economic Impacts of Extended Fisheries Jurisdiction*, ed. Lee Anderson. Ann Arbor, MI: Ann Arbor Science Publishers

Burke, Leslie, D. Liew, M. Etter, C. Annand, G. Peacock, R.O. Boyle, L. Brander, and R. Barbara, Department of Fisheries and Oceans (1994) 'The Scotia-Fundy Inshore Dragger Fleet ITQ Program.' Notes for presentation to American Fisheries Society, 23 August

Byron, R.F. (1976) 'Economic Functions of Kinship Values in Family Businesses: Fishing Crews in North Atlantic Communities.' *Sociology and Social Research* 60, 147–60

Cameron, Silver Donald (1990) 'Net Losses: The Sorry State of Our Atlantic Fishery.' *Canadian Geographic*, April/May, 29–37

Carter, Roger (1982) *Something's Fishy: Public Policy and Private Corporations in the Newfoundland Fishery.* St John's, Nfld.: St John's Oxfam Committee

Chandler, Alfred D., Jr (1962) *Strategy and Structure.* Cambridge, MA: MIT Press

Chandler, Alfred D., Jr, and Herman Daems, eds. (1980) *Managerial Hierarchies.* Cambridge, MA: Harvard University Press

Charles, Anthony T. (1988) 'Fishery Socioeconomics: A Survey.' *Land Economics* 64, 276–95

– (1992) 'Canadian Fisheries: Paradigms and Policy.' In *Canadian Ocean Law and Policy*, ed. David VanderZwaag. Toronto: Butterworths

Charles River Associates, Inc. (1982) 'Gulf of Maine Adjudication: Government Policy Study and Impact Analysis.' Report to the U.S. Department of State, June

Cheticamp Development Commission (1991) 'Cheticamp, Nova Scotia: An Analysis of the Impact Caused by a Downturn in the Fishing Industry.' March

Clark, C.W. (1976) *Mathematical Bioeconomics: The Optimal Management of Renewable Resources.* New York: Wiley-Interscience

Clement, Wallace (1984) 'Canada's Coastal Fisheries: Formation of Unions, Cooperatives and Associations' *Journal of Canadian Studies* 19, 5–33

188 References

- (1986) *The Struggle to Organize: Resistance in Canada's Fishery*. Toronto: McClelland and Stewart
Commercial Fisheries News (1991) 'New Bedford Holds on to Top-Valved Crown.' May, 21A
- (1992) 'Groundfish Advisors Submit Recommendations.' September, 1A
- (1994) 'Canada's TAGS: Aid, Reduced Fishing Capacity.' July, 19B
Connelly, Patricia, and Martha MacDonald (1983) 'Women's Work: Domestic and Wage Labour in a Nova Scotian Community.' *Studies in Political Economy* 10, 45–62
- 'Women and Rural Economic Development: Case Studies from Nova Scotia.' Mimeo, St Mary's University
Copes, Parzival (1982) 'Implementing Canada's Marine Fisheries Policy.' *Marine Policy* 6, 219–35
- (1983) 'Fisheries Management on Canada's Atlantic Coast: Economic Factors and Socio-Political Constraints.' *Canadian Journal of Regional Science* 6, 1–32
Corey, R., and J. Dirlam (1982) 'Canadian Financial Assistance to the Fishing Industry.' Center for Ocean Management Studies, University of Rhode Island, NOAA/Sea Grant Marine Memorandum 73, September
Crowley, R.W., and H. Palsson (1992) 'Rights Based Fisheries Management in Canada.' *Marine Resource Economics* 7, 1–22
Crutchfield, James A. (1979) 'Economic and Social Implications of the Main Policy Alternatives for Controlling Fishing Effort.' *Journal of the Fisheries Research Board of Canada* 36, 742–52
Crutchfield, S.R. (1985) 'The Impact of Groundfish Imports on the U.S. Fishing Industry: An Empirical Analysis.' *Canadian Journal of Agricultural Economics* 33, 195–207
Crutchfield, S.R., and J.M. Gates (1984) 'Measuring the Effect on Fisheries of the 200-Mile Zone.' *Maritimes*, 6–7 November
Davis, Anthony, and Richard L. Kasdan (1984) 'Bankrupt Government Policy and Belligerent Fishermen's Responses: Dependency and Conflict in Southwest Nova Scotia.' *Journal of Canadian Studies* 19, 108–124
Davis, Anthony, and Victor Thiessen (1986) 'Making Sense of the Dollars: Income Distribution among Atlantic Canadian Fishermen and Public Policy.' *Marine Policy* 10, 201–14
- (1988) 'Public Policy and Social Control in the Atlantic Fisheries.' *Canadian Public Policy* 14, 66–77
Demsetz, Harold (1988) *The Organization of Economic Activity*. Oxford: Basil Blackwell
Department of Fisheries and Oceans, Canada (DFO) (various years) *Canadian Fisheries Annual Statistical Review*. Ottawa Statistics and Analysis Group, Economic Policy Branch, Economic Development Directorate

- (various years) *Canadian Fisheries Landings*, 1978–1990
- Economic Analysis Division, Program Coordination and Economics Branch, Scotia-Fundy Region (1991) *1990 Report on Fishing Vessel Performance: Scotia Fundy Region*, Economic and Commercial Analysis Report No. 103. Ottawa: Supply and Services Canada, August
- (1992) 'Northern Cod Moratorium.' *Backgrounder*, July
- (1993a) *Fisheries Management: A Proposal for Reforming Licensing, Allocation and Sanctions Systems*. Ottawa: Supply and Services Canada
- (1993b) *Report on the Status of Groundfish Stocks in the Canadian Northwest Atlantic.*' July
- (1993c) *Charting a New Course: Towards the Fishery of the Future*. Report of the Task Force on Incomes and Adjustments in the Atlantic Fisheries, November
- (1994) *Atlantic Groundfish Management Plan*. Ottawa: Supply and Services Canada
Department of Regional Industrial Expansion, Canada (1976) *Community and Employment Implications of Restructuring the Atlantic Groundfisheries*
Dewar, Margaret (1983) *Industry in Trouble: The Federal Government and the New England Fisheries* (Philadelphia, PA: Temple University Press
- (1986) 'The New England Fishing Industry.' In Doeringer, Moss, and Terkla (1986a)
Digou, D. (1992) *Scotia-Fundy Region Harvesting Sector Overview, 1986–1991*, Economic and Commercial Analysis Report No. 122. Ottawa: Supply and Services Canada, NS, July
Doeringer, Peter B. (1984) 'Internal Labor Markets and Paternalism in Rural Areas.' In *Internal Labor Markets*, ed. Paul Osterman. Cambridge: MIT Press
Doeringer, Peter B., Philip I. Moss, and David G. Terkla (1986a) *The New England Fishing Economy: Jobs, Income, and Kinship*. Amherst, MA: University of Massachusetts Press
- (1986b) 'Capitalism and Kinship: Do Institutions Matter in the Labor Market?' *Industrial and Labor Relations Review* 40, 48–60
Edwards, Steven F., and S.A. Murawski (1993) 'Potential Economic Benefits from Efficient Harvest of New England Groundfish.' *North American Journal of Fisheries Management* 13, 437–49
Elbaum, Bernard (1986) 'The Steel Industry before World War I.' In *The Decline of the British Economy*, ed Bernard Elbaum and William Lazonick. Oxford: Oxford University Press
Elbaum, Bernard, and William Lazonick (1986) 'An Institutional Perspective on British Decline.' In *The Decline of the British Economy*, ed. Bernard Elbaum and William Lazonick. Oxford: Oxford University Press
Employment and Immigration, Canada (1991) *Unemployment Insurance and Fishing*. Ottawa: Supply and Services Canada

Environment Canada (1976) *A Policy for Canada's Commercial Fisheries*. Ottawa: Queen's Printer

Feeny, D., F. Berkes, B.J. McKay, and J.M. Acheson (1990) 'The Tragedy of the Commons: Twenty-two Years Later.' *Human Ecology* 18, 1–19

Felixson, T., P.G. Allen, and D.A. Storey (1994) 'An Econometric Model of the Market for Fresh New England Groundfish with Emphasis on the Role of Canadian Imports.' *Northeast Journal of Agricultural and Resource Economics* 35, 24–34

Ferris, J.S., and C.G. Plourde (1982) 'Labour Mobility, Seasonal Unemployment Insurance, and the Newfoundland Inshore Fishery.' *Canadian Journal of Economics* 15, 426–41

Fisheries Resource Conservation Council (FRCC) (1993a) *1993 Conservation Requirements for Atlantic Groundfish*, Report to the Minister of Fisheries and Oceans. Ottawa: Supply and Services Canada. 23 August

– (1993b) *1994 Conservation Requirements for Atlantic Groundfish*, Report to the Minister of Fisheries and Oceans. Ottawa: Supply and Services Canada, November

Fishery Products International (FPI) (1991) *Annual Report*

Food Marketing Institute (1990) *Food Marketing Industry Speaks, 1990*. Washington, DC: Food Marketing Institute

– (1991) *Trends 91*. Washington, DC: Food Marketing Institute

Foreman, Christopher H. (1982) 'Sea of Troubles: Managing New England's Fisheries.' *Regulation*, July/August 43–7

Fritts, Charles (1980) 'Stifling the Fishery.' Letter from the Legislative Counsel of the New Bedford Seafood Council. *New York Times*, 10 July

Gardner, Michael (1988) 'Enterprise Allocation System in the Offshore Groundfish Sector in Atlantic Canada' *Marine Resource Economics* 5, 389–454

Gardner Pinfold Consulting Economists Ltd (1986) 'The Fishermen's Unemployment Insurance Program.' Mimeo prepared for the Royal Commission on Unemployment Insurance, August

Georgianna, Daniel L., and Joel Dirlam (1982) *Industrial Structure and Cost of Fresh Atlantic Groundfish Processing*, Final Report for Contract 80-FA-C-0045 to National Marine Fisheries Service, November

Georgianna, Daniel L., J. Dirlam, and R. Townsend (1993) *The Groundfish and Scallop Processing Sectors in New England*, Final Report: USDOC, Contract #50EANF-2-00065, July

Georgianna, Daniel L., and William. V. Hogan (1986) 'Production Costs in Atlantic Fresh Fish Processing.' *Marine Resource Economics* 2, 275–92

Georgianna, Daniel L., and R. Ibara (1983) 'Groundfish Processing in Massachusetts during the 1970s.' *Marine Fisheries Review* 45, 1–10

Giasson, Marie (1992) 'Capital and Work-force Adaptation in Clare.' In Apostle and Barrett (1992a)

Haché, J.E. (chairman) (1989) *Report of the Scotia–Fundy Groundfish Task Force.* Ottawa: Supply and Services Canada for the Department of Fisheries and Oceans, December

Hall, Peter (1990) 'Crisis in the Atlantic Fishery.' *Canadian Business Review* 17, 44–8

Halliday, R.G., F.G. Peacock, and D.L. Burke (1992) 'Development of Management Measures for the Groundfish Fishery in Atlantic Canada: A Case Study of the Nova Scotia Inshore Fleet.' *Marine Policy* 16, 411–26

Hasselback, N., and K. Marris (1991) 'Seafood on the Upswing.' *Seafood Business,* July–August, 38–48

Heaney, William H. (1979) 'Innovative Enterprise and Social Change: A Case Study of Resource Control and Social Transactions in a Quebec Maritime Community.' In Raoul Andersen, ed. (1979a)

Hennessey, Timothy M., and Michael LeBlanc (1985) 'Fishery Administration and Management in Canada and the United States: Implications for Georges Bank.' In *Georges Bank: A Book and Atlas,* ed. R.H. Backus. Cambridge, MA: MIT Press

Hogan, W., and D. Georgianna (1989) 'U.S. Fish Processing Capacity and Imports of Whole Groundfish from Canada.' *Marine Resource Economics* 6, 213–25

Hogan, William, Daniel Georgianna, and Toby Huff (1991) *The Massachusetts Marine Economy.* Dartmouth, MA: Massachusetts Centers of Excellence Corporation, University of Massachusetts Dartmouth

Holmsen, Andreas A. (1972) 'Remuneration, Ownership, and Investment Decisions in the Fishing Industry.' University of Rhode Island Marine Technical Report Number 1, mimeo, Kingston, RI, January

H.R. 5557 (1992) 'To Amend the Magnuson Fishery Conservation and Management Act to Provide for the Restoration of New England Stocks of Groundfish and for Other Purposes.' 102D Congress, 2D Session, 2 July

Ilcan, Susan (1985) 'The Position of Women in the Nova Scotia Secondary Fishing Industry: A Community-Based Study.' Working Paper No. 8–86, Gorsebrook Research Institute For Atlantic Canada Studies, St Mary's University, Halifax

Innis, Harold A. (1940) *The Cod Fisheries.* New Haven, CT: Yale University Press

– (1954) *The Cod Fisheries: The History of an International Economy.* Toronto: University of Toronto Press

Jenkins, J.T., (1921) *A History of the Whale Fisheries.* Port Washington, NY: Kennikat Press

Johnson, R.N., and G.D. Libecap (1982) 'Contracting Problems and Regulations: The Case of the Fishery.' *American Economic Review* 72, 1005–22

Jorion, Paul (1982) 'All-Brother Crews in the North Atlantic.' *Canadian Review of Sociology and Anthropology* 19, 513–26

Katz, Larry L. (1986) 'Efficiency Wage Theories: A Partial Evaluation.' In *National Bureau of Economic Research Macroeconomic Annual*, ed. Stanley Fisher. Cambridge, MA: MIT Press

Kimber, Stephen (1989) *Net Profits: The Story of National Sea*. Halifax: Nimbus Publishing

Kirby, Michael (1982) *Navigating Troubled Waters: A New Policy for the Atlantic Fisheries*. Report of the Task Force on Atlantic Fisheries. Ottawa: Supply and Services Canada, December

Krueger, Anne O. (1974) 'The Political Economy of the Rent-Seeking Society.' *American Economic Review* 64, 291–303

Krugman, Paul R. (1989) 'Industrial Organization and International Trade.' In *Handbook of Industrial Organization*, ed. Richard Schmalensee and Robert Willig. Amsterdam: North-Holland

Lamson, Cynthia (1986) 'On the Line: Women and Fish Plant Jobs in Atlantic Canada.' *Relations Industriel* 41, 145–56

Lamson, Cynthia, and Arthur J. Hanson, eds (1984) *Atlantic Fisheries and Coastal Communities: Fisheries Decision-Making Case Studies*. Halifax: Dalhousie Ocean Studies Programme, Dalhousie University

Lazear, Edward P. (1991) 'Labor Economics and the Psychology of Organizations.' *Economic Perspectives* 5, 89–110

Lorenz, Edward, and Frank Wilkinson (1986) 'The Shipbuilding Industry, 1880–1965.' In Elbaum and Lazonick (1986)

MacDonald, David W. (1980) *Power Begins at the Cod End: The Newfoundland Trawlerman's Strike, 1974–1975*. Social and Economic Studies No. 26, Institute of Social and Economic Research, Memorial University of Newfoundland, St John's, Nfld.

MacDonald, Martha, and M. Patricia Connelly (1986a) 'A Leaner Meaner Industry: A Case Study of "Restructuring" in the Nova Scotia Fishery.' Mimeo, International Working Seminar on Social Research and Public Policy Formation in the Fisheries, Institute of Fisheries, University of Tromso, Tromso, Norway, 14–21 June

– (1986b) 'A Cadillac Plant: Restructuring the Labour Process in Nova Scotia Fish Plants.' Mimeo, Canadian Political Science Association Annual Meetings, Winnipeg, Manitoba, 6 June

MacDonald, R.D.S. (1984) 'Canadian Fisheries Policy and the Development of Atlantic Coast Groundfisheries Management.' In Lamson and Hanson, eds (1984)

Magnuson Fishery Conservation and Management Act of 1976, as amended by PL 95-354, PL 96-61, and PL 96-561, 31 December 1980, Section 3, #18

Mandale, M., and J.F. Morley (1990) 'The Atlantic Fishery in the 1990s: Background to Crisis.' *Atlantic Report* 25: 2, Atlantic Provinces Economic Council, Halifax, NS

Martin, Kent O. (1979) 'Play by the Rules or Don't Play At All: Space Division and Resource Allocation in a Rural Newfoundland Fishing Community.' In Andersen, ed. (1979a)

Massachusetts Division of Employment Security (DES) (1986) Unpublished employment data

Massachusetts Offshore Groundfish Task Force (1990) *New England Groundfish in Crisis – Again.* Publication No. 16,551-42-200-1-91–CR, December

Matthews, David R. (1976) *'There's No Better Place Than Here': Social Change in Three Newfoundland Communities.* Toronto: Peter Martin Associates Ltd

– (1993) *Controlling Common Property: Regulating Canada's East Coast Fishery.* Toronto: University of Toronto Press

Mazany, R.L., L.G. Barrett, and R.A. Apostle (1987) 'Market Segmentation: Nova Scotia Fish Processing and the U.S. Market.' *Marine Policy* 11, 29–44

Meaney, John (1992) 'Federal Fisheries Law and Policy: Controls on the Harvesting Sector.' In *Canadian Ocean Law and Policy*, ed. David VanderZwaag. Toronto: Butterworths

Miller, Marc L., and John Van Maanen (1979) 'Boats Don't Fish, People Do: Some Ethnographic Notes on the Federal Management of Fisheries in Gloucester.' *Human Organization* 38, 377–85

Ministry of External Affairs, Canada (1983) *A Review of Canadian Trade Policy.* Ottawa: Supply and Services Canada

Ministry of State for Economic Development, Canada (1984) *Assistance to Business in Canada (ABC).* Ottawa: Supply and Services Canada

National Sea Products Ltd and Fishermen, Food, and Allied Workers (1991) 'Agreement'

New England Fishery Management Council (NEFMC) (1993) 'Final Supplemental Environmental Impact Statement for Amendment #5 to the Northeast Multispecies Fishery Management Plan.' Saugus, MA, 30 September

Nova Scotia Department of Fisheries (1977) *Sea, Salt and Sweat: A Story of Nova Scotia and the Vast Atlantic Fisheries.* (Halifax: Nova Scotia Department of Fisheries

Olsen, Mancur (1988) 'The Productivity Slowdown, the Oil Shocks, and the Real Cycle.' *Journal of Economic Perspectives* 2: 4, 43–71

Opaluch, James J., and Nancy E. Bockstael (1984) 'Behavioral Modeling and Fisheries Management.' *Marine Resource Economics* 1, 105–15

194   References

Orbach, Michael (1977) *Hunters, Seamen and Entrepreneurs* Berkeley: University of California Press
– (1978) 'Social and Cultural Aspects of Limited Entry.' In *Limited Entry as a Fishery Management Tool*, ed. R.B. Rettig and J.C. Ginter. Seattle: University of Washington Press
Paine, R. (1978) 'That Outport Culture: Review Essay.' *Canadian Review of Sociology and Anthropology* 25, 148–56
Pearse, P.H. (1981) 'Fishing Rights, Regulations, and Revenue.' *Marine Policy* 5, 135–46
Peters, T.J., and R.H. Waterman, Jr. (1982) *In Search of Excellence*, New York: Harper and Row
Peterson, Susan B., and D.L. Georgianna (1988) 'New Bedford's Fish Auction: A Study in Auction Method and Market Power.' *Human Organization* 47, 235–41
Pinkerton, E.W., ed. (1989) *Co-Operative Management of Local Fisheries: New Directions for Improved Management and Community Development*. Vancouver: University of British Columbia Press
Plante, Janice M. (1992) 'Groundfish Advisors Submit Recommendations.' *Commercial Fisheries News*, September, 1A, 9A
– (1994) 'Scientists Report: Georges Bank Cod, Yellowtail Stocks at Critical Juncture.' *Commercial Fisheries News*, September, 1A, 18A
Poggie, John, and Richard Pollnac (1978) 'Social Desirability of Work and Management among Fishermen in Two New England Ports.' Anthropology Working Paper No. 5, International Center for Marine Resource Development, Kingston, RI
Pollnac, Richard B., Carl Gersuny, and John J. Poggie, Jr (n.d.) 'Economic Gratification Patterns among Fishermen and Millworkers in Southern New England.' Mimeo, Department of Sociology-Anthropology, University of Rhode Island
Pollnac, Richard, and John Poggie (1988) 'The Structure of Job Satisfaction among New England Fishermen and Its Application to Fisheries Management Policy.' *American Anthropologist* 90, 888–901
Porter, Michael E. (1980) *Competitive Strategy*. New York: The Free Press
Portland Fish Exchange (1994) *Price and Landings Annual 1992 and 1993*, sixth edition, Portland, ME
Proskie, John, and J.C. Adams (1969) *1969 Survey of the Labour Force in the Offshore Fishing Fleet, Atlantic Coast*. Ottawa: Economics Branch, Fisheries Service, Department of Fisheries and Forestry
Pross, A.P., with W. Heber (1982) 'The Social and Economic Impact of the Lunenburg Fishing Industry.' Mimeo, Economic Research Unit, Department of Fisheries and Oceans, Ottawa, April

Quiggin, John (1988) 'Private and Common Property Rights in the Economics of the Environment.' *Journal of Economic Issues* 22, 1071–87

Raymond, Janice L., ed. (1985) *Scotia-Fundy Region Fishing Community Profiles*, Canadian Data Report of Fisheries and Aquatic Sciences No. 540, Ottawa: Department of Fisheries and Oceans, Scotia-Fundy Region, October

Ruddle, K., E. Hviding, and R.E. Johannes (1992) 'Marine Resources Management in the Context of Customary Tenure.' *Marine Resource Economics* 7, 249–73

Safer, Andrew (1994a) 'Northern Cod Collapse Devastates Newfoundland.' *Commercial Fisheries News*, June, 14–16B

– (1994b) 'What Went Wrong: Science, Politics, and Nature.' *Commercial Fisheries News*, July, 8–9B, 19B

Santopietro, George D. and L.A. Shabman (1992) 'Can Privatization Be Inefficient?: The Case of the Chesapeake Bay Oyster Fishery.' *Journal of Economic Issues* 26, 407–19

Schrank, W.E., E. Tsoa, and N. Roy (1987) 'The Canadian–United States Trade in Fish and Fish Products.' In *Resource Economies in Emerging Free Trade, Proceedings of a Maine/Canadian Trade Conference*. Orono, ME: University of Maine Press

Simon, Herbert A. (1991) 'Organizations and Markets.' *Economic Perspectives* 5: 25–44

Sinclair, Peter R. (1983) 'Fishermen Divided: The Impact of Limited Entry Licensing in Northwest Newfoundland.' *Human Organization* 42, 307–13

– (1984) 'Fishermen of Northwest Newfoundland: Domestic Commodity Production in Advanced Capitalism.' *Journal of Canadian Studies* 19, 34–48

– (1986) 'The Survival of Small Capital: State Policy and the Dragger Fleet in North-West Newfoundland.' *Marine Policy* 10, 111–18

– (1987) *State Intervention and the Newfoundland Fisheries*. Brookfield, VT: Gower Publishing Company

Smith, Leah J., and Susan B. Peterson (1977) *The New England Fishing Industry: A Basis For Management*, Technical Report No. WHOI-77-57, Woods Hole Oceanographic Institution, August

Squires, Susan E. (1990) 'Kin and Religion: Organizational Changes in Crew Recruitment in the Fishing Industry of Newfoundland, Canada.' Unpublished doctoral dissertation, Department of Anthropology, Boston University

Statistics Canada (various years) *Exports by Commodity*. Ottawa: Supply and Services Canada

Steinberg, Charles (1974) 'The Legal Problems in Collective Bargaining by Canadian Fishermen.' *Labor Law Journal* 25, 643–54

Stiglitz, Joseph E. (1991) 'Symposium on Organizations and Economics.' *Economic Perspectives* 5: 2, 15–24

Stiles, Geoffrey (1979) 'Labor Recruitment and the Family Crew in Newfoundland.' In Andersen, ed. (1979a)

Sutinen, Jon G. (1979) 'Fishermen's Remuneration Systems and Implications for Fisheries Development.' *Scottish Journal of Political Economy* 26:2, 147–62

Swaney, James (1990) 'Common Property Reciprocity, and Community.' *Journal of Economic Issues* 24, 451–62

Taylor, Roger (1992) 'National Sea Products To Downsize.' *Chronicle-Herald and Mail-Star*, Halifax, NS, 7 May, B6

Terkla, David G., Peter B. Doeringer, and Philip I. Moss (1988) 'Widespread Labor Stickiness in the New England Offshore Fishing Industry: Implications for Adjustment and Regulation.' *Land Economics* 4, 73–82

Terkla, David G., and J. Wiggin (1994) 'Gloucester Waterfront Study: Land Use and Economics.' Appendix 5 of the Special Resource Study for Gloucester, MA, North Atlantic Regional Office, National Park Service, Boston, MA

Thiessen, Victor, and Anthony Davis (1988) 'Recruitment to Small Boat Fishing and Public Policy in the Atlantic Canadian Fisheries.' *Canadian Review of Sociology and Anthropology* 25, 603–27

Townsend, Ralph (1990) 'Entry Restrictions in the Fishery: A Survey of the Evidence. *Land Economics* 66, 359–78

Tsoa, E., W.E. Schrank, and N. Roy (1982) 'U.S. Demand for Selected Groundfish Products 1967–1980.' *American Journal of Agricultural Economics* 64, 483–9

U.S. Bureau of the Census (1991–2) *Imports for Consumption.* Washington, DC: USGPO

– (1993) *Imports of Merchandise*

U.S. Department of Commerce (USDOC), National Marine Fisheries Service (various years) *Fisheries of the United States.* Washington, DC: USGPO

– (1986a) Unpublished data

– (1986b) 'NOAA Fishery Management Study,' 30 June

– (1986c) 'Final Affirmative Countervailing Duty Determination: Certain Fresh Atlantic Groundfish from Canada.' *Federal Register* 51, 10041–69. International Trade Administration, 24 March

– (1990) Unpublished data

– (1991a) 'Status of the Fishery Resources off the Northeastern United States for 1990.' Woods Hole, MA: Northeast Fisheries Center, January

– (1991b) 'Preliminary 1990 Landings and Values of New England Fish and Shellfish.' Northeast Fisheries Center News Release, April

– (1994) 'Status of the Fishery Resources off the Northeastern United States for 1993.' Woods Hole, MA: Northeast Fisheries Center, January

U.S. Department of Labor (USDOL) (various years) *Employment and Wages, Annual Averages*

U.S. Fish and Wildlife Service (1992) Interview, August

U.S. International Trade Commission (USITC) (1984) *Conditions of Competition Affecting the Northeastern U.S. Groundfish and Scallop Industries in Selected Markets,* Washington, DC: USITC Publication 1622, December

– (1986) *Certain Fresh Atlantic Groundfish from Canada.* Washington, DC: USITC Publication 1844, May

Van Maanen, John, Marc L. Miller, and Jeffrey C. Johnson (n.d.) 'An Occupation in Transition: Traditional and Modern Forms of Commercial Fishing.' Mimeo

VanderZwaag, David (1983) 'Canadian Fisheries Management: A Legal and Administrative Overview.' *Ocean Development and International Law Journal* 13, 171–209

Warner, William (1983) *Distant Water: The Fate of the North Atlantic Fisherman.* Toronto: Little, Brown

Weber, Max, (1964) *The Theory of Social and Economic Organizations.* New York: Free Press

Weeks, E., and L. Mazany, (1983) *The Future of the Atlantic Fisheries.* Montreal: Institute for Research on Public Policy

White, Donald J. (1954) *The New England Fishing Industry.* Cambridge, MA: Harvard University Press

Williamson, Oliver E. (1975) *Markets and Hierarchies.* New York: The Free Press

Wilson, James A. (1980) 'Adaptation to Uncertainty and Small Numbers Exchange: The New England Fresh Fish Market.' *Bell Journal Of Economics* 11, 491–504

– (1982) 'The Economical Management of Multispecies Fisheries.' *Land Economics* 58, 417–34

Wilson, James, P. Klebon, S. McKay, and R. Townsend (1991) 'Chaotic Dynamics in a Multiple Species Fishery: A Model of Community Predation.' *Ecological Modeling* 58, 303–22

– (1992) 'When Is Luddism Efficient?' *Marine Resource Economics* 7: 2, 86–91

Author Index

Subject Index